# Immunochemistry of the Extracellular Matrix

# Volume II Applications

Editor
**Heinz Furthmayr, M.D.**
Associate Professor
Department of Pathology
Yale University
New Haven, Conneticut

**CRC Press, Inc.**
**Boca Raton, Florida**

**Library of Congress Cataloging in Publications Data**

Main entry under title:

Immunochemistry of the extracellular matrix.

Bibliography: p.
Includes index.
1. Collagen. 2. Immunochemistry.
3. Ground substance (Anatomy) 4. Extra-
cellular space. I. Furthmayr, Heinz.
II. Title: Extracellular matrix.
QP552.C6I45 616.7′9 81-18062
ISBN 0-8493-6196-6 (v. 1) AACR2
ISBN 0-8493-6197-4 (v. 2)



Direct all inquiries to CRC Press, 2000 Corporate Blvd., N.W., Boca Raton, Florida, 33431.



International Standard Book Number 0-8493-6196-6 (Volume I)
International Standard Book Number 0-8493-6197-4 (Volume II)

Library of Congress Card Number 81-18062
Printed in the United States

# PREFACE

Fibrous connective tissue is the single most prevalent tissue in the body and in a variety of forms, such as bone, tendon, cartilage, and fascia, it gives structural rigidity. Collagen, as the major protein constituent of connective tissue is found in addition throughout the body forming structures such as basement membranes, and it is part of the extracellular matrix to which cells adhere to form specific organs. Five genetically distinct types of collagen with distinct physicochemical and structural properties have been described thus far. These collagens are synthesized by various mesenchymal and epithelial cells and apparently are distributed in tissues and organs in specific ways. In a variety of pathological conditions expression, distribution and control of production and/or degradation are affected.

The main purpose of the two volumes on *Immunochemistry of the Extracellular Matrix* is to describe state of the art methods, which have been proven to provide antibody reagents of defined specificity to collagens as well as other glycoproteins found in association within connective tissue. The isolation and purification of collagens, procollagens, and related connective tissue proteins are described for several tissues, tissue culture cells and species. Immunization with these collagens in laboratory animals yields antibodies with different characterization specificities: to the procollagen extension fragments, the non-helical segments of the α-chains, to helical and denatured determinants. Methods are described to obtain, isolate, and characterize these antibody specificities by selection of the appropriate animal for immunization, immunoadsorbtion procedures, and sensitive serological assays. Experimental data are given and previous results are discussed in terms of producing antibody reagents of desired specificity. Also included are immunohistological methods at the level of the right microscope and the electron microscope aimed at elucidating the localization of these collagens in vivo in a variety of tissues.

Connective tissue biochemistry, biology, and pathology are rapidly expanding fields today and the information assembled in these volumes should encourage and facilitate the use of immunochemical tools for studies in these important areas of research. Finally, the authors tried to define several challenging areas of inquiry in connective tissue biology in which specific antibodies can contribute to our understanding of cellular processes.

The task of assembling these two volumes was greatly facilitated by the enthusiasm of the contributing authors, all experts in the field, who regarded this project as a worthwhile and timely effort and I am grateful for their support, effort, and valuable contributions.

In writing my own contributions I came to realize more than ever, how much I owe to my teachers Rupert Timpl and Klaus Kühn, and to the many friends I have found amongst the group of connective tissue biochemists. Amongst the individuals at Yale I particularly wish to thank Joseph Roll and Joseph Madri, who not only helped to bring this project to fruition, but who were also responsible for initiating connective tissue reseach in the Department of Pathology together with me.

I welcome comments and criticisms from readers.

November 1981

Heinz Furthmayr

## THE EDITOR

**Heinz Furthmayr, M.D.,** University of Vienna, Austria, is Associate Professor of Pathology at Yale University. He received his training in immunochemistry in the Department of Immunochemistry in Munich, Germany, and in membrane biochemistry at the Department of Pathology at Yale University. He has been a Visiting Fellow in the Department of Experimental Biology and Genetics at the CSSR Academy of Sciences in Prague and at the Hôpital St. Louis in Paris. In 1978 he received a Research Career Award from the American Cancer Society. His publications include over 70 scientific articles, several reviews and chapters on connective tissue proteins and plasma membrane glycoproteins. He is a member of the American Society of Immunologists and the American Society of Pathologists and serves on Advisory Committees of the U.S. Public Health Service.

## CONTRIBUTORS

### Volume II

**John Bateman, Ph.D.**
Visiting Fellow
National Cancer Institute
National Institutes of Health
Bethesda, Maryland

**Mario Chojkier, M.D.**
Consultant
National Cancer Institute
National Institutes of Health
Bethesda, Maryland

**Heinz Furthmayr, M.D.**
Associate Professor
Department of Pathology
Yale University
New Haven, Connecticut

**Gary R. Grotendorst, Ph.D.**
Postdoctoral Fellow
National Institute of Dental Research
National Institutes of Health
Bethesda, Maryland

**A. Tyl Hewitt, Ph.D.**
NIH Expert
National Institute of Dental Research
National Institutes of Health
Bethesda, Maryland

**John D. Kemp, M.D.**
Postdoctoral Fellow in Immunology and Pathology
Department of Pathology
Yale University
New Haven, Connecticut

**Hynda K. Kleinman, Ph.D.**
Research Chemist
National Institute of Dental Research
National Institutes of Health
Bethesda, Maryland

**Joseph A. Madri, M.D., Ph.D.**
Assistant Professor
Department of Pathology
Yale University
New Haven, Connecticut

**George R. Martin, Ph.D.**
Chief, Laboratory of Developmental Biology and Anomalies
National Institute of Dental Research
National Institutes of Health
Bethesda, Maryland

**Klaus von der Mark, Ph.D.**
Head of Research Group
Max-Planck-Institute for Biochemistry
Martinsreid, West Germany

**Bjorn R. Olsen, Ph.D.**
Professor of Biochemistry
College of Medicine and Dentistry
Rutgers Medical School
Piscataway, New Jersey

**Donna M. Pesciotta**
Research Associate
Department of Anatomy
Albert Einstein College of Medicine
Bronx, New York

**Beverly Peterkofsky, Ph.D.**
Research Biochemist
National Cancer Institute
National Institutes of Health
Bethesda, Maryland

**David H. Rohrbach, Ph.D.**
Postdoctoral Fellow
National Institute of Dental Research
National Institutes of Health
Bethesda, Maryland

**F. Joseph Roll, M.D.**
Assistant Professor
Department of Medicine
Liver Center
University of California
San Francisco, California

**H. E. J. Seppä, M.D.**
Professor of Anatomy
Oulu Medical School
Oulu, Finland

**Elliott Schiffman, Ph.D.**
Research Chemist
National Institute of Dental Research
National Institutes of Health
Bethesda, Maryland

**Victor P. Terranova, D.D.S., Ph.D.**
Staff Fellow
National Institute of Dental Research
National Institutes of Health
Bethesda, Maryland

**Rupert Timpl, Ph.D.**
Head of Research Group
Max-Planck-Institute for Biochemistry
Martinsreid, West Germany

**Hugh H. Varner, Ph.D.**
Postdoctoral Fellow
National Institute of Dental Research
National Institutes of Health
Bethesda, Maryland

**Charlotte M. Wilkes**
Medical Student
Eastern V.A. Medical College
Virginia Beach, Virginia

# TABLE OF CONTENTS

## Volume I

## Volume II

Chapter 1

# THE CELL BIOLOGY OF COLLAGEN SECRETION

**D. M. Pesciotta and B. R. Olsen**

## TABLE OF CONTENTS

## I. INTRODUCTION

The general pathway for synthesis and intracellular transport of proteins that are destined for secretion from cells is now firmly established based on studies with a variety of systems.[1] Secretory proteins are synthesized on membrane-bound ribosomes and translocated across the membrane of the rough endoplasmic reticulum (RER). The binding of the nascent polypeptides to the membrane of the RER and the subsequent cotranslational transfer across the membrane is probably mediated by a hydrophobic signal sequence.[2–4] These signal sequences are in most cases found at the amino termini of the nascent polypeptides, and they are rapidly removed from the polypeptides during the cotranslational transfer process.[2,3] After being segregated in the lumen of the RER, the secretory polypeptides are packaged and transported by a vesicular mechanism to the Golgi complex.[1,5] Once in the Golgi complex, the secretory products are concentrated and packaged into secretory granules or vacuoles for transport to the surface of the cell.[1,5,6] While in the RER and the Golgi complex, secretory proteins are modified by a number of co- as well as posttranslational processes. These include core-glycosylation of glycoproteins in RER via the dolicholphosphate pathway,[7,8] hydroxylations,[9–11] rapid cleavage of signal peptides as well as limited proteolysis,[2–4,12–14] and trimming of carbohydrate side chains with addition of terminal sugar residues.[7,15,16] After extensive modification, the secretory products finally leave the cell by exocytosis.[1,5,6]

Initial studies of connective tissue cells appeared to support the view that synthesis and secretion of collagen differed from the general pathway outlined above for other secretory proteins. Instead, it was suggested that collagen may be directly transported from ribosomes through the cytoplasm to the cell surface without the involvement of vesicles and extruded into the extracellular matrix through the cell membrane. Based on several studies from different laboratories,[17–22] in part with antibodies to procollagen[23–26] and prolyl hydroxylase,[25] it is now clear that secretion of collagen follows the same general pathway as secretion of other proteins.

The procollagen polypeptides are first synthesized on ribosomes of the RER and transferred across the membrane into the lumen of the RER. This is followed by concentration and packaging of the procollagen molecules in the Golgi complex and finally secretion by exocytosis. During the intracellular transport, the procollagen polypeptides are extensively modified. Thus, the basic cellular mechanisms responsible for synthesis and secretion of procollagen are not unique for cells synthesizing procollagen, but are shared by all cells producing proteins for export. However, the size, rigidity, and relative insolubility of collagen may have required collagen-producing cells to develop specific modifications of these basic secretory processes. For this reason, detailed studies on collagen synthesis and secretion may well provide additional insight into the general process of protein secretion.

## II. STRUCTURE OF PROCOLLAGENS

As discussed in detail in Chapter 1, this volume, collagens comprise a family of protein molecules which share certain structural features. Each collagen molecule contains three α-chains arranged in a triple helical conformation. Each α-chain contains about 1000 amino acid residues, and with the exception of short sequences (telopeptides) at the ends of the chain, every third amino acid residues in the chain is glycine.

Procollagen (see Figure 1) is the biosynthetic precursor of collagen[27] and is the molecular form that apparently is discharged into the extracellular matrix by cells

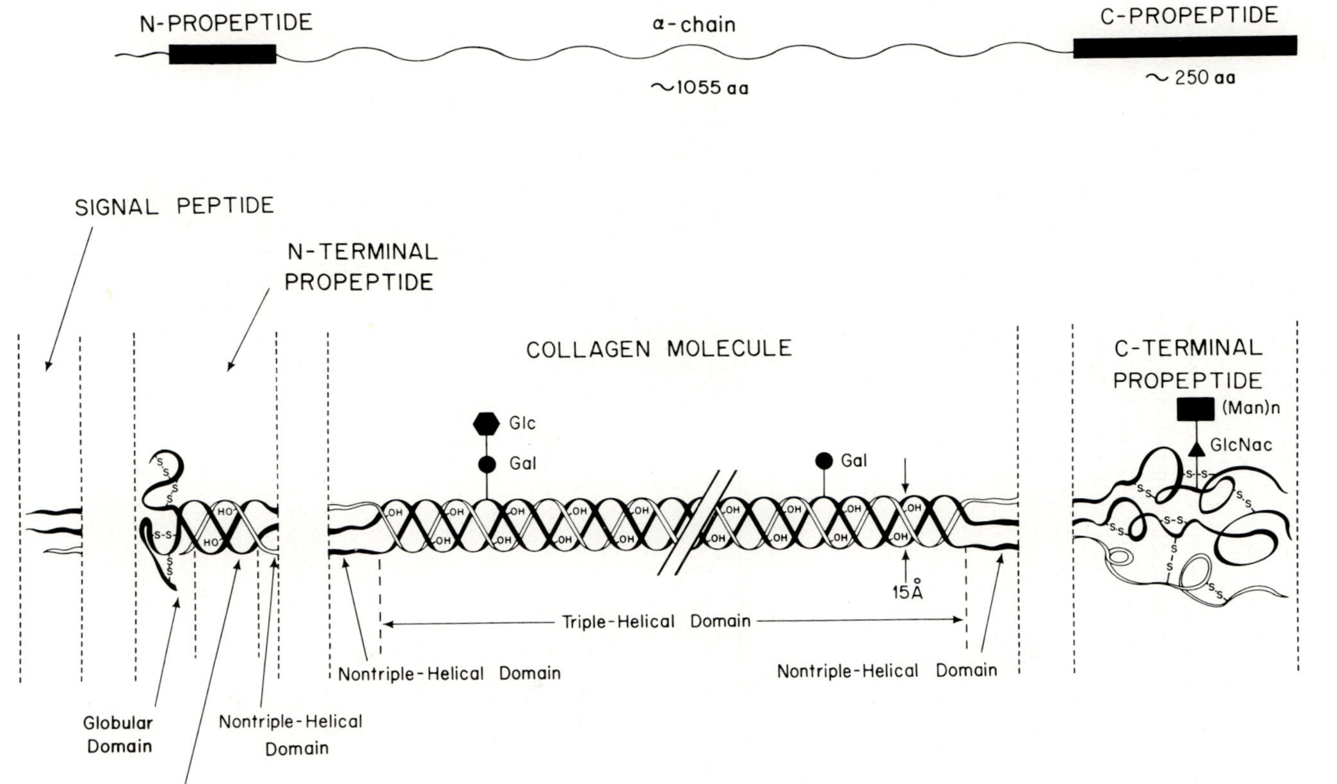


FIGURE 1. Diagram showing the structure of preproα-chains and procollagen Type I. Symbols: Glc (glucose); Gal (galactose); Man (mannose), and GlcNac (N-acetylglucosamine). (From Olsen, B. R., *Cell Biology of the Extracellular Matrix*, Hay, E. D., Ed., Plenum Press, 1981, 139. With permission.)

synthesizing collagen. Procollagen molecules differ from collagen molecules in that each of the three α-chains contain polypeptide extensions at both the amino and carboxyl ends. These propeptides are removed from procollagen by specific procollagen N- and C-proteases during the conversion to collagen in the extracellular matrix.[27]

The amino propeptide of proα1(I)-chains has been well characterized from a variety of tissues.[28-32] These studies have indicated that the propeptide is organized into three structural domains and contains several intrachain disulfide bonds. At the amino terminus of the peptide is a globular domain of about 100 amino acid residues. The central domain contains about 45 amino acid residues in a collagen-like triple-helical structure. This region is linked to the collagen α-chains by a short connecting peptide that is not in a triple-helical conformation. The amino propeptide of proα2(I)-chains is considerably smaller than the amino propeptide of proα1(I)-chains and is largely composed of collagen-like sequences.[27]

In comparison, relatively little is known about the carboxyl propeptides. Only the carboxyl propeptide of chick Type I procollagen has been studied in some detail.[31,33-35] These studies have shown that the carboxyl propeptide of Type I procollagen is a trimer consisting of two different subunits in a 2 to 1 ratio linked together by interchain disulfide bonds, each subunit contains intrachain disulfide bonds and an asparagine-linked oligosaccharide sidechain. Carbohydrate analysis of this oligosaccharide side chain has revealed the presence of 2 residues of N-acetyl glucosamine and 9 to 13 residues of mannose within each subunit.[33,34]

Studies on the physiological processing of the amino and carboxyl propeptides have indicated that these are removed "*en bloc*" in a stepwide fashion from the procollagen molecule.[36-38] For Type I procollagen, it appears that the amino propeptides are removed first, followed by removal of the carboxyl propeptide. This may not be a general pathway for all types of procollagens, however. For example, in smooth muscle cell cultures, the carboxyl propeptide may be removed before the amino propeptide,[39] and in Type III procollagen, the amino propeptide may be cleaved only to a limited extent.[40,41]

## III. COLLAGEN POLYPEPTIDES ARE SYNTHESIZED AS PREPROα-CHAINS ON MEMBRANE-BOUND POLYSOMES

Initial biochemical studies designed to localize the site of procollagen synthesis in collagen-producing cells were conducted by incubating cells or tissues with radioactive proline and then analyzing isolated subcellular fractions for the radioactivity. Many of these experiments suggested that collagen polypeptides were synthesized on membrane-bound polysomes,[17,42,43] but the technical difficulties in preparing homogeneous subcellular fractions caused many investigators to interpret the data with caution.

Autoradiographic studies have also been employed to localize intracellular sites of collagen synthesis while simultaneously tracing its secretion pathway. The labeling pattern found in odontoblasts from rats injected with [3H]-proline[22] showed that the radioactivity was localized in the RER and its associated ribosomes. This indicated that collagen polypeptides were being synthesized on membrane-bound ribosomes. However, even these experiments were not conclusive because they lacked specificity since proline could be incorporated into a variety of proteins in addition to collagen.

Use of specific antibodies against procollagen (see Figure 2) and prolyl hydroxylase (see Figure 3) provided direct evidence for the microsomal location of pro-

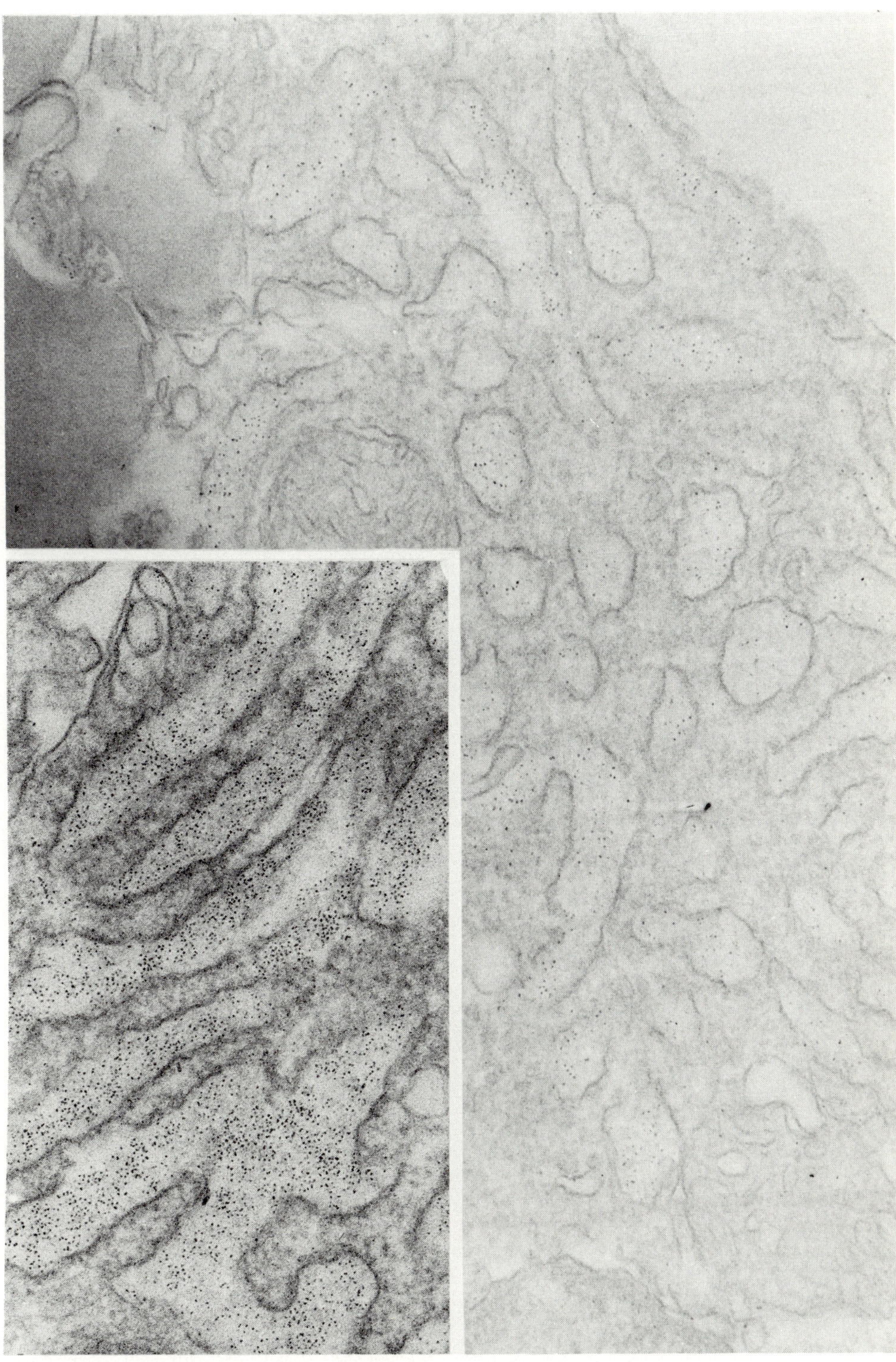

FIGURE 2. Electron micrograph showing portions of chick tendon fibroblasts incubated with ferritin-labeled antibodies directed against Type I procollagen. For a description of the method used for immunolabeling, see Chapter 3, Volume II and Olsen et al.[25] Note the labeling of the RER.

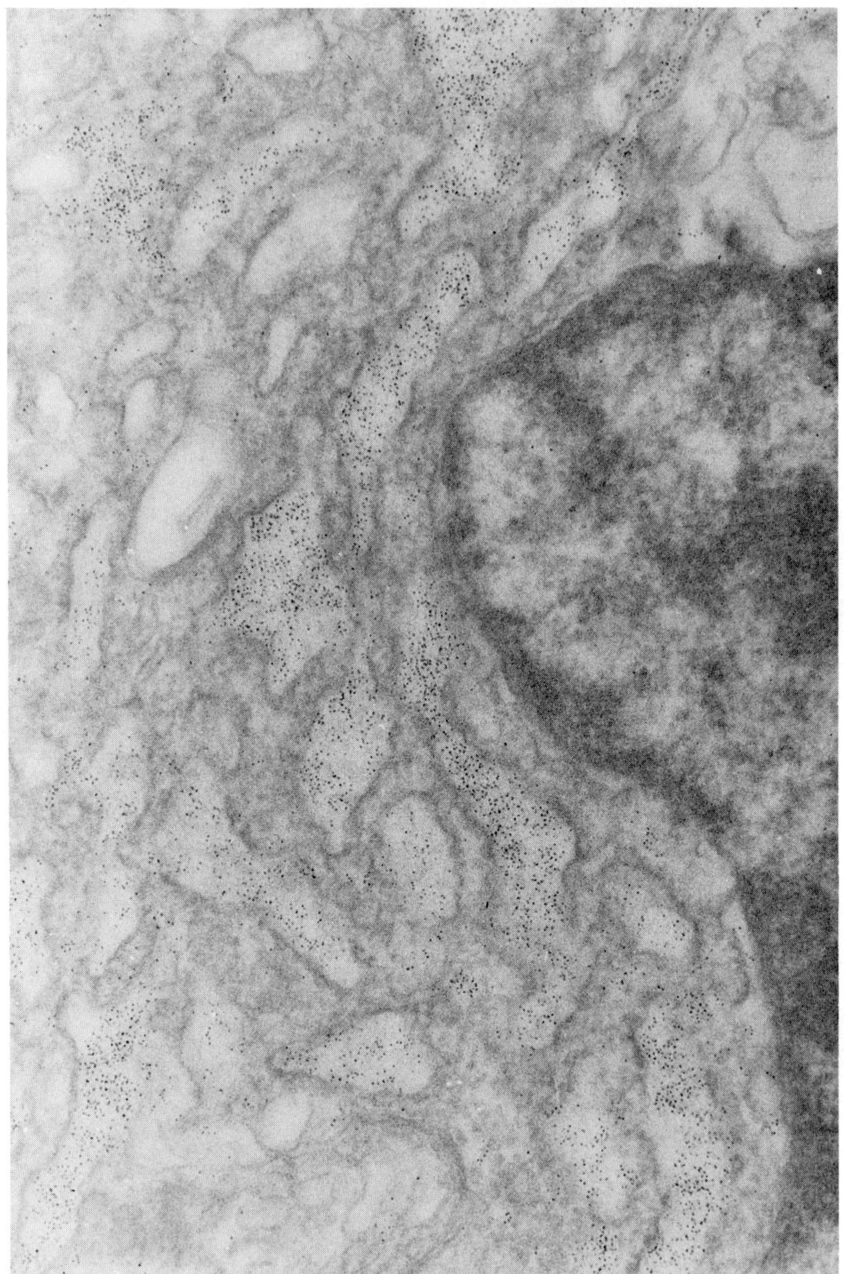

FIGURE 3. Portion of chick tendon fibroblast labeled with ferritin-conjugated antibodies directed against prolyl hydroxylase showing labeling of the cisternae of the RER. (From Olsen, B., Berg, R., Kishida, Y., and Prockop, D. J., *J. Cell Biol.*, 64, 340, 1975. With permission.)

collagen synthesis. Ferritin-labeled antibodies directed against the carboxyl propeptides of Type I procollagen[23,24] and against prolyl hydroxylase[25] were employed to localize procollagen and one of its modification enzymes in fibroblasts isolated from chick embryo tendons. The labeling pattern clearly showed the presence of procollagen and prolyl hydroxylase in the RER of the fibroblasts. In addition, im-

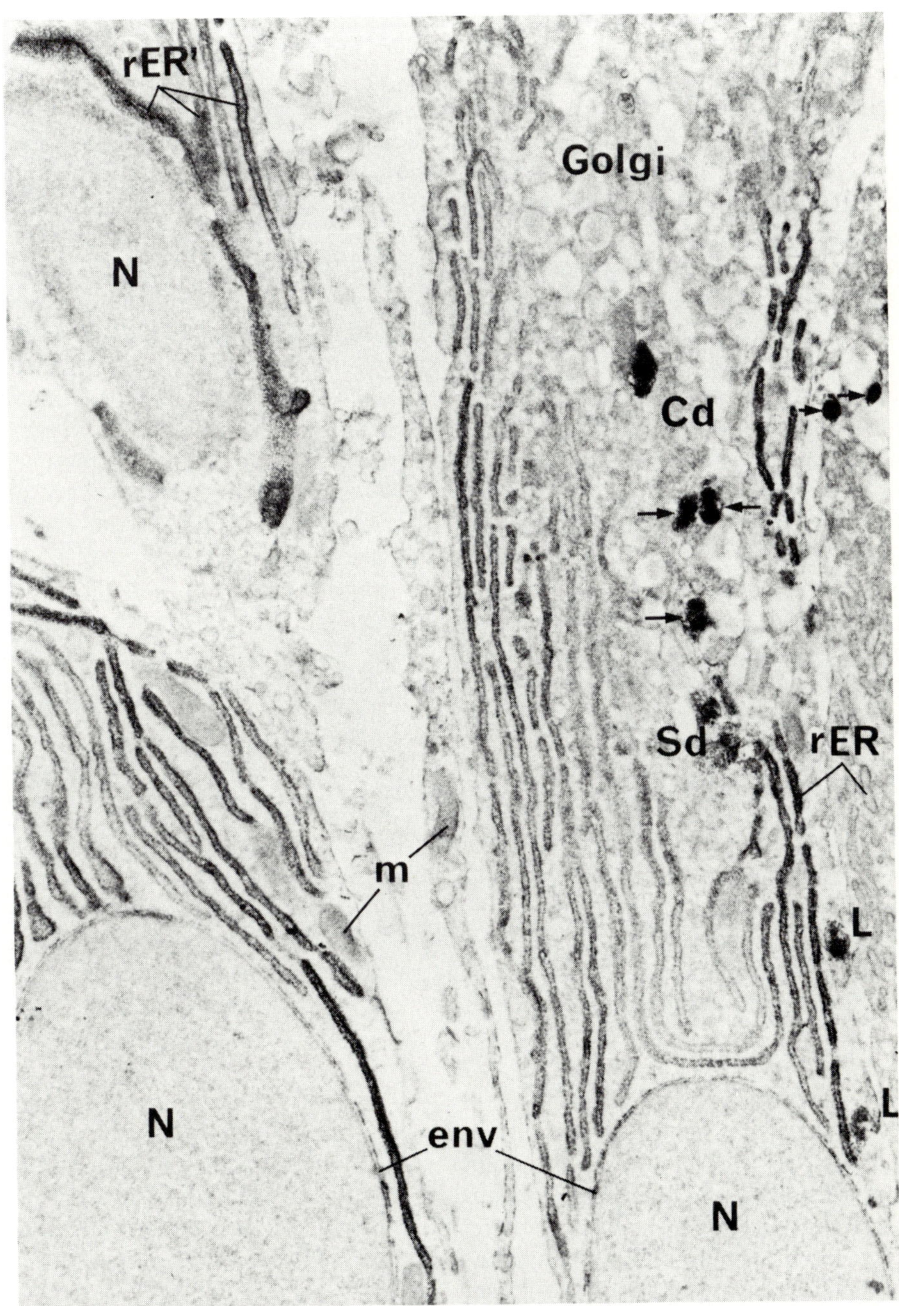


FIGURE 4. Longitudinal section of odontoblasts immunostained by exposure to peroxidase-conjugated antibodies directed against Type I procollagen followed by incubation with diaminobenzidine-$H_2O_2$ to detect sites of peroxidase activity. Note labeling of RER (rER) and limited reactions within the nuclear envelopes (env) of nuclei (N). The Golgi region of the cell in the right half of the figure shows spherical distensions with reaction product (Sd) and cylindrical distentions, one of which shows reaction product (Cd). Several secretory granules (arrows) show intense labeling. (From Karim, A., Cournil, I., and Leblond, C. P., *J. Histochem. Cytochem.*, 27, 1070, 1979. With permission.)

munocytochemical techniques with peroxidase-labeled antibodies have been employed to locate the proα-chains within the organelles of odontoblasts and in predentin.[26] These studies have verified what was previously seen in cultured fibroblasts—immunostaining within the RER due to the presence of Type I proα-chains (see Figure 4). The labeling pattern also indicated procollagen within the Golgi complex. This observation plus the demonstration of procollagen in the cisternae of the RER and the microsomal location of prolyl hydroxylase virtually rules out the possibility that procollagen secretion does not involve the classical RER-Golgi complex pathway.

Recent studies on translation of collagen mRNA in cell-free systems[44] have shown that proα-chains are initially synthesized as larger chains referred to as preproα-chains[45,47] (Figure 1). Sequencing studies have indicated that these preproα-chains contain hydrophobic signal sequences at their amino-termini.[46,47] This illustrates that procollagen polypeptides like other nascent secretory polypeptides contain a hydrophobic region necessary for transmembrane transfer in the RER. This hydrophobic region is believed to participate in forming a membrane ribosome junction that will enable the nascent chain to be translocated to the cisternal space of the RER. During or after translocation of the nascent polypeptide chain, the signal peptide is usually proteolytically removed by a membrane-bound protease.[2–4] Detailed studies on the processing of the signal peptides in preproα-chains have not been reported, but preliminary data suggest that the signal sequences are rapidly removed during or shortly after ribosomal synthesis of preproαchains.[46–48] Thus, proα-chains isolated from microsomes do not contain amino-terminal signal sequences.

## IV. COTRANSLATIONAL AND POSTTRANSLATIONAL MODIFICATIONS OF INTRACELLULAR PROCOLLAGEN

Upon entering the cisternal space of the RER, the nascent proα-chains are extensively modified. These modifications include *N*- and *O*-glycosylation, hydroxylation, and possibly the formation of disulfide bonds.[10,11] Several types of data suggest that by the time procollagen leaves the RER, it is hydroxylated, glycosylated, disulfide-bonded, and perhaps at least in part, folded into the correct conformation.[10] However, the functional significance of these modifications and the order in which they occur is only partially understood.

### A. Hydroxylation of Procollagen

Hydroxylations of prolyl and lysyl residues by three hydroxylases located in the RER are the best characterized modifications of procollagen.[10,49,50] The three hydroxylases convert some prolyl residues to 4-hydroxyproline or 3-hydroxyproline and some lysyl residues to hydroxylysine. The hydroxylases are mixed-function oxygenases, and the reactions involve molecular oxygen, α-ketoglutarate, ferrous iron, and ascorbate. During hydroxylation of the peptide substrate, α-ketoglutarate is decarboxylated to succinate. The three hydroxylases require a peptide substrate in a nontriple-helical conformation, and the susceptible prolyl and lysyl residues must be within specific sequences.[49] Thus, prolyl 4-hydroxylase and lysyl hydroxylase will only hydroxylate prolyl and lysyl residues in the Y position of peptides containing X-Y-Gly-sequences, whereas prolyl 3-hydroxylase will only hydroxylate prolyl residues in the X position of peptides containing X-Hyp-Gly-sequences.[49]

Prolyl 4-hydroxylase has been purified from several sources.[49] The enzyme is a tetramer of molecular weight of 240,000 daltons with the subunit structure $\alpha_2\beta_2$, where α and β are two types of subunits of molecular weight of about 60,000 daltons.

The enzyme is located within the RER, as shown by electron microscopy using ferritin-labeled antibodies against the chick embryo enzyme[25] (see Figure 3) and by its presence in microsomal subcellular fractions.[17,19,21] In freshly isolated chick tendon fibroblasts, most of the immunoreactive enzyme appears to be in an active tetrameric form of molecular weight of 240,000 daltons.[51] However, several studies indicate that from about 30 to 99 percent of the enzyme protein reacting with specific antibodies is in an inactive form of molecular weight of 60,000 daltons in cells and tissues that synthesize collagen.[49,52,54] Comparisons of various cells and tissues indicate that the ratio of active tetramers to the inactive form is proportional to the rate of collagen synthesis. It is possible, therefore, that conversion of prolyl hydroxylase from an inactive to an active form plays a role in regulating collagen synthesis.[55] Prolyl 4-hydroxylase does not appear to be an integral membrane protein or tightly associated with the membrane of the RER since it is easily soluble after mild detergent treatment of microsomes to disrupt the microsomal membrane.[21] In contrast, lysyl hydroxylase, also a microsomal enzyme,[19,49] appears to be tightly associated with the membrane of the RER. Immunocytochemical localization of lysyl hydroxylase within the RER has not been reported. Presumably, lysyl hydroxylation is occurring simultaneously with prolyl hydroxylation on nascent proα-chains, but may continue (with prolyl hydroxylation) after the proα-chains have been released from ribosomes.[10]

What is the function of hydroxylation of prolyl and lysyl residues in collagen? The role of 4-hydroxyproline is to stabilize the triple-helical conformation of collagen. The triple-helix of collagen is not stable at 37°C unless about 90 residues of 4-hydroxyproline are present per α-chain.[10,49] This fact and the fact that prolyl 4-hydroxylase does not hydroxylate peptides in a triple-helical conformation defines an upper limit for the hydroxylation process in the cell: hydroxylation of prolyl residues stops when the molecule folds into a triple-helical conformation.

The function of hydroxylysine in collagen is illustrated by the Type VI of the Ehlers-Danlos syndrome.[56,57] In this form of the syndrome, the predominant symptoms are skin and joint hyperextensibility due to decreased tensile strength of collagen fibers. The biochemical basis for the defect appears to be a defect in lysyl hydroxylase leading to synthesis of hydroxylysine-deficient collagen.[57]

### B. Glycosylation of Procollagen

Subsequent to hydroxylation of lysyl residues, some of the hydroxylysine residues become glycosylated. The formation of *O*-β-D-galactosyl-hydroxylysyl and 1-*O*-α-D-glucosyl-*O*-β-D-galactosyl-hydroxylysyl residues is catalyzed by the two enzymes galactosyl transferase and glucosyl transferase. As is the case with hydroxylation of lysyl and prolyl residues, glycosylation of hydroxylysyl residues is initiated while the proα-chains are still attached to the ribosomes. Supporting data for this conclusion come from studies demonstrating the presence of galactosyl-hydroxylysyl and glucosylgalactosyl-hydroxylysyl residues in nascent polypeptide chains and the presence of the two transferase activities in microsomes.[58,60] As the glucosyl transferase does not act on triple-helical substrates,[58] hydroxylysyl glycosylation must be completed within the cisternae of the RER before triple-helix formation occurs.

In addition to glycosylated hydroxylysyl residues, procollagen polypeptides also contain carbohydrate chains linked to asparagine.[34,61–64] In Type I procollagen, the carbohydrate chain is located within the carboxyl propeptide and contains two residues of *N*-acetyl-glucosamine and about ten residues of mannose per proα-chain.[33,34] The available data suggest that the addition of the mannose-rich carbohydrate side chain to the asparagine residue occurs via a lipid-linked intermediate and requires an Asn-X-Thr/Ser sequence within the polypeptide chain.[65,66] In fact, recent sequence analysis of the carboxyl propeptide from Type I procollagen has shown that the carbohydrate attachment site in the proα1(I)-chain is -Ala-Thr-Gln-Asn-Val-Thr-Tyr-His- (see Figure 5). For the proα2-chain the corresponding sequence is -Ala-Ser-Gln-Asn-Ile-Thr-Tyr-His-.[64] The location of the *N*-glycosylation event within the fibroblast has not been identified, but available data from other secretory systems suggest that this process occurs on the cisternal side of the RER membrane.[7,8] Evidence from other proteins also suggests that the mannose-rich core when transferred to the protein acceptor contains glucose residues on its outer branches.[7] These glucose residues are removed from the core, while the polypeptide chains are still in the lumen of the RER. It has been suggested that the removal of these glucose residues may be necessary for the transport of secretory proteins from the RER to the Golgi complex.[67]

### C. Formation of Interchain Disulfide Bonds and Folding of Procollagen Polypeptides into Triple-Helical Structures

Folding of procollagen molecules into a triple-helical conformation is probably preceded by the formation of disulfide bonds between the proα-chains.[10] Type I and II procollagen contain interchain disulfide bonds within the carboxyl propeptides, while Type III procollagen has been shown to contain interchain disulfide bonds between the amino propeptides as well.[27] There is evidence that the interchain disulfide bonds in procollagen form after translation.[68] This is supported by experiments employing α,α′-dipyridyl to inhibit hydroxylation and triple-helix formation in chick embryo tendon fibroblasts. Ferritin-labeled antibodies directed against the carboxyl propeptide of Type I procollagen showed heavy labeling of the RER in both control cells and in cells incubated with α,α′-dipyridyl.[69] Analysis of procollagen polypeptides contained within subcellular fractions isolated from such cells indicated that proα-chains within the microsomes were nontriple-helical, but nevertheless linked by interchain disulfide bonds in cells incubated with α,α′-dipyridyl.[69]

The presence of proα-chains linked by interchain disulfide bonds within the lumen of the RER indicates that these bonds form within this organelle. It has been suggested that triple-helix formation also must occur in the RER.[10] Much of the evidence for this is based on the observed retention of nonhelical disulfide linked proα-chains within the lumen of the RER when cells are incubated under conditions that prevent triple-helix formation because of inhibition of prolyl and lysyl hydroxylation in proα-chains.[10] Isolated proα-chains are compact, flexible structures that presumably are easily packaged into small transport vesicles.

Once proα-chains associate into triple-helical structures, they form rigid rod-like structures over 300 nm long. If the triple-helix indeed forms within the RER prior to transport of the procollagen molecules to the Golgi complex, one is therefore faced with the problem of packaging a rigid rod-like molecule over 300 nm long into transport vesicles that are about 50 nm in diameter. In fact, if triple-helix formation occurs before procollagen molecules are transported from the RER to the Golgi complex, small, 50-nm transport vesicles are probably not involved in the translocation of procollagen from RER to the Golgi complex.

For this reason, studies aimed at locating the site of triple-helix formation are of considerable interest from a cell-biological standpoint. In spite of a large number of such studies, however, the answer to this question appears quite elusive. Our ignorance about the time and the cellular location of the folding of procollagen molecules is largely due to the inadequate methodology available. Physical chemical techniques to measure conformation cannot be used with crude cellular extracts,

```
FIGURE 5.
         1   2          144 145 146 147 148 149 150 151 152 153 154      243 244 245 246
C1:     Asp-Asp-------Ala-Thr-Gln-Asn-Val-Thr-Tyr-His-Cys-Lys-Asn-----Val-Cys-Phe-Leu
                                   |
                                  CHO
         1   2          143 144 145 146 147 148 149 150 151 152 153      242 243 244 245
C2:     Asp-Gln-------Ala-Ser-Gln-Asn-Ile-Thr-Tyr-His-Cys-Lys-Asn-----Val-Cys-Phe-Lys
                                   |
                                  CHO
```

FIGURE 5. Partial amino acid sequences of the proα1 (I) and proα2 (I) carboxyl propeptides showing the amino-terminal sequences of the propeptides and the attachment sites for the mannose-rich oligosaccharide side chains (CHO) in the peptides. (See Pesciotta D. M., Dickson, L. A., Showalter, A. M., Eikenberry, E. F., de Crombrugghe, B., Fietzek, P. P., and Olsen, B. R., *FEBS Lett.*, 125, 170, 1981.)

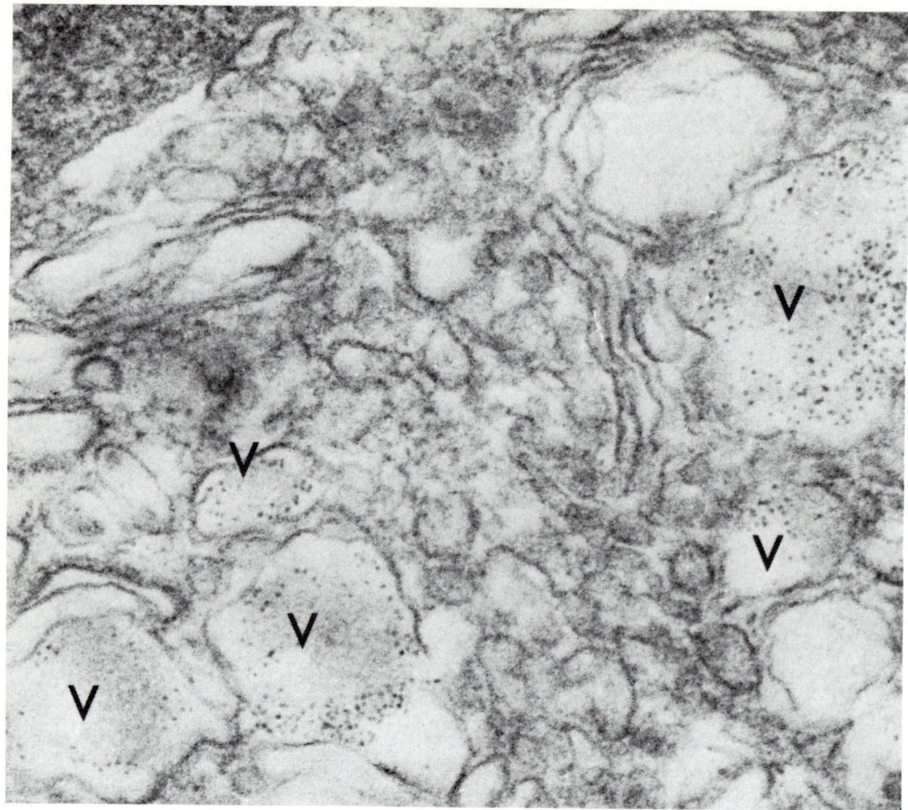


FIGURE 6. Golgi region of tendon fibroblast labeled with ferritin-conjugated antibodies directed against Type I procollagen. Many vacuoles (v1 to 7) are labled in this micrograph. Electron-dense contents may be completely (vac) or partially (v1) lost. Arrows indicate breaks in Golgi vacuoles; nu, nucleus; dc, dilated cisterna. The insert shows a secretory granule, at a higher magnification. (From Nist, C., Hay, E. D., Olsen, B. R., Bornstein, P., Ross, R., and Dehm, P., *J. Cell Biol.*, 65, 75, 1975. With permission.)

and the use of indirect methods to study conformation such as proteolytic enzyme probes makes it difficult to draw firm conclusions. The use of antibodies directed against conformational antigenic determinants in collagen triple-helices should be explored in future studies.

## V. THE ROLE OF THE GOLGI COMPLEX IN PROCOLLAGEN PACKAGING AND SECRETION

The intracellular transport of proα-chains from the cisternae of the RER to the Golgi complex is believed to be mediated via small transport vesicles (but see above). If procollagen synthesis follows a pattern similar to that of the pancreatic acinar cell, these vesicles would fuse with the membranes of the Golgi complex and discharge their contents into the cisternal space of the Golgi elements.

That procollagen is contained within Golgi elements was demonstrated by the use of immunoferritin techniques with antibodies against procollagen (see Figure 6).[23,24] Indirect evidence for this secretory pathway was provided by autoradiographic studies of odontoblasts from rats injected with ($^3$H)-proline.[22] Within 20 min after injection, the ($^3$H)-proline was present not only in the Golgi saccules, but also in

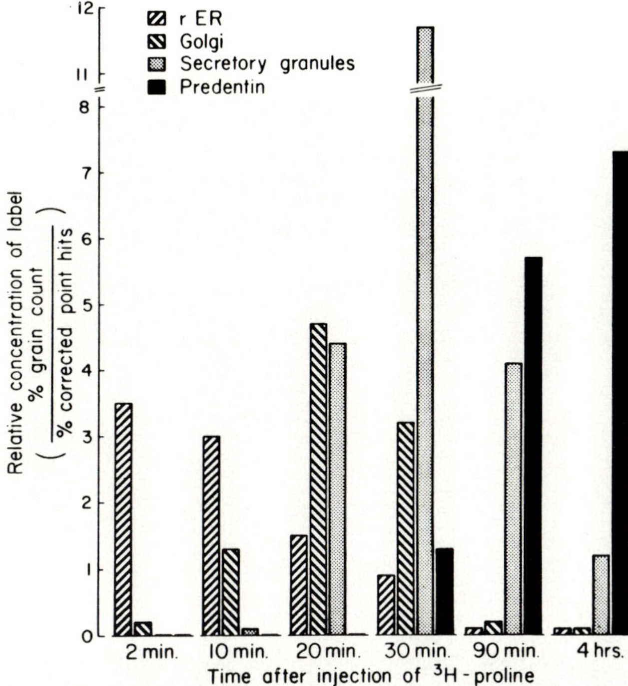


FIGURE 7. Graphic representation of silver grain counts over odontoblast organelles at various times after injection of [$^3$H] proline into rats. After perfusion of animals with paraformaldehyde, sections of incisor teeth were processed for electron microscopical autoradiography. The grain counts are expressed as an index of the relative concentration of label over the RER (rER), Golgi, secretory granules, and predentin as determined by dividing the percent silver grains over each structure by the percent corrected point hits made over each structure. (From Weinstock, M. and Leblond, C. P., *J. Cell Biol.*, 60, 92, 1974. With permission.)

prosecretory and secretory granules. Quantitative analysis revealed an increase in radioactivity in certain organelles with time and indicated the passage of label from RER to the Golgi complex and finally into secretory granules (see Figure 7).[20]

More recent experiments have used antibodies against Type I procollagen and immunoperoxidase techniques to locate procollagen at every stage of the secretory process.[26] Antigenic material was found in cisternae of the RER, intermediate or transfer vesicles of the Golgi region, spherical and cylindrical distensions of Golgi saccules, and in secretory granules.[26] Within spherical distensions of the Golgi saccules the antigenic material appeared to be loose entangled threads (see Figure 8).[26] However, in cylindrical distensions, the material appeared as densely packed parallel threads. This has been interpreted to imply that triple-helix formation occurs in the Golgi complex of odontoblasts.[26]

The final stage in the intracellular transport of procollagen involves packaging into secretory granules for transport to the cell surface. Relatively little is known about this process. However, on the basis of morphological data, it has been suggested that the secretion of procollagen includes the release of ordered aggregates of procollagen molecules[26,70,71] and that such aggregates may represent an intermediate in the formation of collagen fibrils in the extracellular matrix.[71]

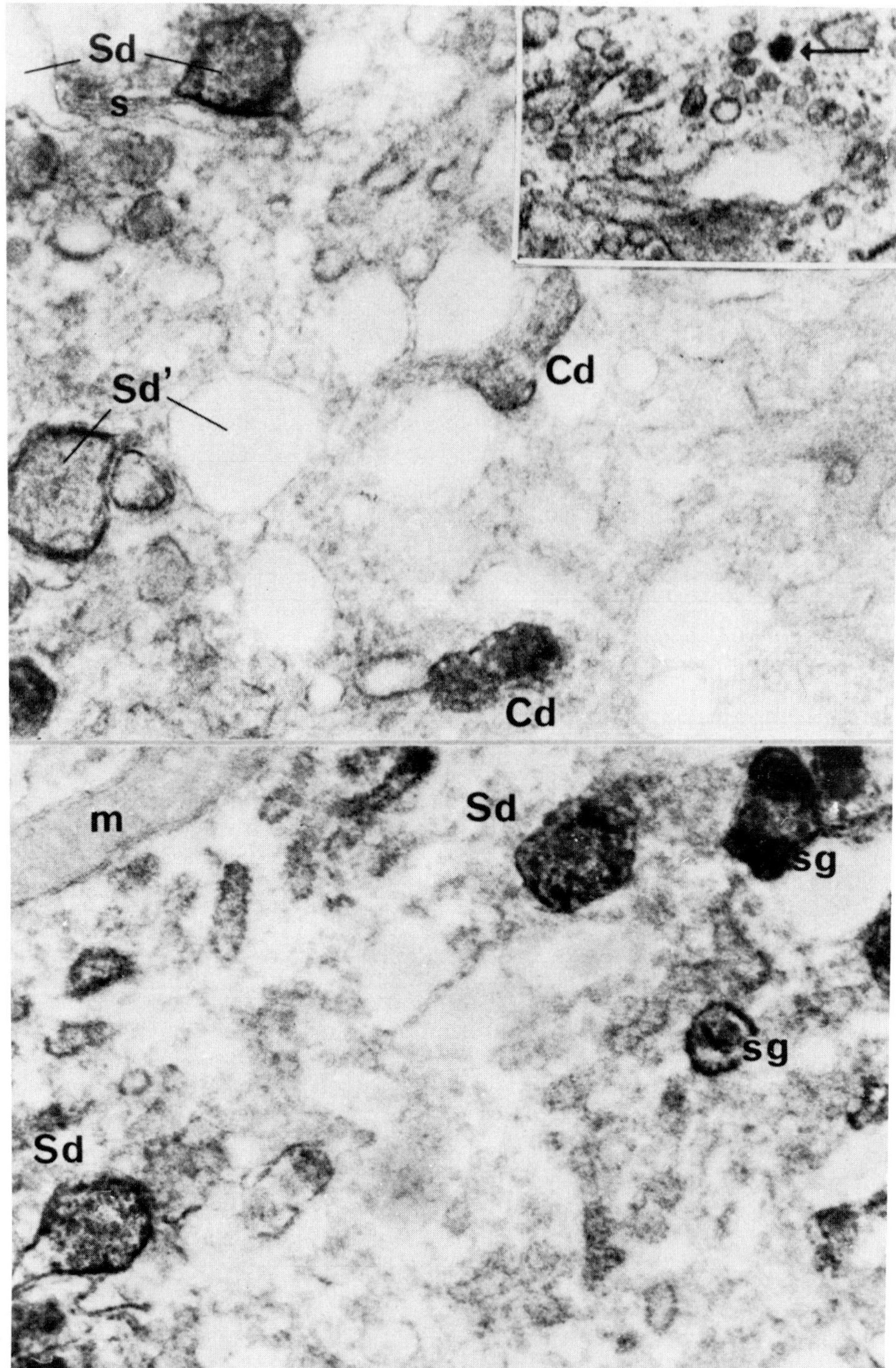


FIGURE 8. Rat odontoblasts stained with peroxidase-conjugated antibodies directed against Type I procollagen. Note staining of spherical distension (Sd) and cylindrical distensions (Cd) of Golgi saccules. Secretory granules (Sg) are also stained. The inset at upper right shows a spherical distension (Sd) associated with transfer vesicles, one of which is intensely stained (arrow). Mitochondria (m) are unstained. (From Karim, A., Cournil, I., and Leblond, C. P., *J. Histochem. Cytochem.*, 27, 1070, 1979. With permission.)

# VI. SUMMARY AND CONCLUDING REMARKS

The synthesis, intracellular transport, and packaging of procollagen represents a unique variation on the general theme of protein secretion. Unique to the procollagen system is the link between molecular assembly and folding of polypeptides and post-translational modifications, such as hydroxylations and glycosylations. Also unique to this system is the packaging of long rigid rod-like molecules into ordered aggregates for exocytotic release from cells. The molecules in such aggregates are processed by extracellular procollagen proteases, and the collagen molecules thus formed participate in extracellular fibril assembly. During the past decade, many intriguing questions about procollagen synthesis and intracellular transport have been answered. A large number of problems remain, however. For example, what is the detailed cellular location of lysyl hydroxylase and the collagen glycosyl transferases? Are 50-nm transport vesicles involved in the transport of procollagen from RER to the Golgi complex? Are coated vesicles involved in this transport? In cells that secrete a mixture of collagen types are the different types of molecules packaged into separate secretory granules? Are procollagen aggregates intermediates in extracellular fibril assembly?

To answer many of these as well as additional questions, the experimental approach will have to be "molecular" rather than "statistical." The techniques chosen will have to permit the study of individual molecules and their behavior rather than the statistical properties of a large number of molecules. In future studies, it is certain that techniques involving the use of electron microscopy wll have to play a major role. The precise identification of molecular structures in such studies will have to be based on the use of specific affinity-labeling reagents, including specific antibodies and lectins. For this reason, efforts to obtain antibodies against components involved in the synthesis, transport, and packaging of procollagen provide the basis for future studies on the cellular handling of this important extracellular protein.

# ACKNOWLEDGMENT

Original contributions from this laboratory were supported in part by research grants AM 21471 and AM 16515 from the National Institutes of Health of the U.S. Public Health Service.

# REFERENCES

1. **Palade, G. E.,** Intracellular aspects of the process of protein secretion, *Science,* 189, 347, 1975.
2. **Blobel, G. and Lingappa, V. R.,**Transfer of proteins across intracellular membranes, in *Transport of Macromolecules in Cellular Systems,* Silverstein, S., Ed., Dahlem Konferenzen, Berlin, 1978, 289.
3. **Davis, B. D. and Tai, P. C.,** The mechanism of protein secretion across membranes, *Nature,* 283, 433, 1980.
4. **Blobel, G.,** Intracellular protein topogenesis, *Proc. Natl. Acad. Sci. U.S.A.,* 77, 1496, 1980.
5. **Jamieson, J. D.,** Intracellular transport and discharge of secretory proteins: present status and future perspectives, in *Transport of Macromolecules in Cellular Systems,* Silverstein, S., Ed., Dahlem Konferenzen, Berlin, 1978, 273.

6. **Farquhar, M. G.,** Traffic of products and membranes through the golgi complex, in *Transport of Macromolecules in Cellular Systems,* Silverstein, S. C., Ed., Dahlem Konferenzen, Berlin, 1978, 341.
7. **Parodi, A. J. and Leloir, L. F.,** The role of lipid intermediates in the glycosylation of proteins in the eucaryotic cell, *Biochim. Biophys. Acta,* 559, 1, 1979.
8. **Struck, D. K. and Lennarz, W. J.,** The function of saccharide-lipids in synthesis of glycoproteins, in *The Biochemistry of Glycoproteins and Proteoglycans,* Lennarz, W. J., Ed., Plenum Press, New York, 1980, 35.
9. **Gallop, P. M. and Paz, M. A.,**Posttranslational protein modifications, with special attention to collagen and elastin, *Physiol. Rev.,* 55, 418, 1975.
10. **Prockop, D. J., Kivirikko, K. I., Tuderman, L., and Guzman, N. A.,** The biosynthesis of collagen and its disorders, *N. Engl. J. Med.,* 301, 13, 1979.
11. **Olsen, B. R.,** Collagen biosynthesis, in *Cell Biology of Extracellular Matrix,* Hay, E. D., Ed., Plenum Press, New York, in press, 1981.
12. **Steiner, D. F., Kemmler, W., Tager, H. S., and Peterson, J. D.,** Proteolytic processing in the biosynthesis of insulin and other proteins, *Fed. Proc. Fed. Am. Soc. Exp. Biol.,* 33, 2105, 1974.
13. **Hasilik, A. and Neufeld, E. F.,** Biosynthesis of lysosomal enzymes in fibroblasts. Synthesis as precursors of higher molecular weight, *J. Biol. Chem.,* 255, 4937, 1980.
14. **Hasilik, A. and Neufeld, E. F.,** Biosynthesis of lysosomal enzymes in fibroblasts. Phosphorylation of mannose residues, *J. Biol. Chem.,* 255, 4946, 1980.
15. **Tabas, I. and Kornfeld, S.,** Biosynthetic intermediates of β-glucuronidase contain high mannose oligosaccharides with blocked phosphate residues, *J. Biol. Chem.,* 255, 6633, 1980.
16. **Harpaz, N. and Schachter, H.,** Control of glycoprotein synthesis. Processing of asparagine-linked oligosaccharides by one or more rat liver Golgi α-D-mannosidases dependent on the prior action of UDP-*N*-acetyl glucosamine: α-D-mannoside β2-*N*-acetylglucosaminyl-transferase I, *J. Biol. Chem.,* 255, 4894, 1980.
17. **Diegelmann, R. F., Bernstein, L., and Peterkofsky, B.,** Cell-free collagen synthesis on membrane-bound polysomes of chick embryo connective tissue and the localization of prolyl hydroxylase on the polysome-membrane complex, *J. Biol. Chem.,* 248, 6514, 1973.
18. **Cutroneo, K. R., Guzman, N. A., and Sharawy, M. M.,** Evidence for a subcellular vesicular site of collagen prolyl hydroxylation, *J. Biol. Chem.,* 249, 5989, 1974.
19. **Harwood, R., Grant, M. E., and Jackson, D. S.,** Collagen biosynthesis: characterization of subcellular fractions from embryonic chick fibroblasts and the intracellular localization of protocollagen prolyl and protocollagen lysyl hydroxylases, *Biochem. J.,* 144, 123, 1974.
20. **Revel, J.-P. and Hay, E. D.,** An autoradiographic and electron microscopic study of collagen synthesis in differentiating cartilage, *Z. Zellforsch. Mikrosk. Anat.,* 61, 110, 1963.
21. **Peterkofsky, B. and Assad, R.,** Submicrosomal localization of prolyl hydroxylase from chick embryo limb bone, *J. Biol. Chem.,* 251, 4770, 1976.
22. **Weinstock, M. and Leblond, C. P.,** Synthesis, migration and release of precursor collagen by odontoblasts as visualized by radioautography after $^3$H-proline administration, *J. Cell Biol.,* 60, 92, 1974.
23. **Olsen, B. R. and Prockop, D. J.,** Ferritin conjugated antibodies used for labeling of organelles involved in the cellular synthesis and transport of procollagen, *Proc. Natl. Acad. Sci. U.S.A.,* 71, 2033, 1974.
24. **Nist, C., von der Mark, K., Hay, E. D., Olsen, B. R., Bornstein, P., Ross, R., and Dehm, P.,** Location of procollagen in chick corneal and tendon fibroblasts with ferritin-conjugated antibodies, *J. Cell Biol.,* 65, 75, 1975.
25. **Olsen, B., Berg, R., Kishida, Y., and Prockop, D. J.,** Further characterization of embryonic tendon fibroblasts and the use of immunoferritin techniques to study collagen biosynthesis, *J. Cell Biol.,* 64, 340, 1975.
26. **Karim, A., Cournil, I., and Leblond, C. P.,** Immunochemical localization of procollagens, *J. Histochem. Cytochem.,* 27, 1070, 1979.
27. **Fessler, J. H., and Fessler, L. I.,** Biosynthesis of procollagen, *Annu. Rev. Biochem.,* 47, 129, 1978.
28. **Becker, V., Timpl, R., Helle, O., and Prockop, D. J.,** $NH_2$-terminal extensions on skin collagen from sheep with a genetic defect in conversion of procollagen into collagen, *Biochemistry,* 15, 2853, 1976.
29. **Hörlein, D., Fietzek, P. P., Wachter, E., Lapière, C. M., and Kühn, K.,** Amino acid sequence of the aminoterminal segment of dermatosparactic calf skin procollagen type I, *Eur. J. Biochem.,* 99, 31, 1979.
30. **Rohde, H., Wachter, E., Richter, W. J., Bruckner, P., Helle, O., and Timpl, R.,** Amino acid sequence of the N-terminal noncollagenous segment of dermatosparactic sheep procollagen type I, *Biochem. J.,* 1979, 631, 1979.

31. **Pesciotta, D. M.,** The Primary Structure of the Amino and Carboxyl Propeptides from Type I Procollagen, Ph.D. thesis, Rutgers University, New Brunswick, N.J., 1981.
32. **Pesciotta, D. M., Silkowitz, M. H., Fietzek, P. P., Graves, P. N., Berg, R. A., and Olsen, B. R.,** Purification and characterization of the amino-terminal propeptide of proα1 (I) chains from embryonic chick tendon procollagen, *Biochemistry,* 19, 2447, 1980.
33. **Olsen, B. R., Guzman, N. A., Engel, J., Condit, C., and Aase, S.,** Purification and characterization of a peptide from the carboxy-terminal region of chick tendon procollagen type I, *Biochemistry,* 16, 3030, 1977.
34. **Clark, C. C.,** The distribution and initial characterization of oligosaccharide units on the COOH-terminal propeptide extensions of the proα1 and proα2 chains of type I procollagen, *J. Biol. Chem.,* 254, 10798, 1979.
35. **Showalter, A. M., Pesciotta, D. M., Eikenberry, E. F., Yamamoto, T., Pastan, I., deCrombrugghe, B., Fietzek, P. P., and Olsen, B. R.,** Nucleotide sequence of a collagen cDNA-fragment coding for the carboxyl end of proα1 (I)-chains, *FEBS Lett.,* 111, 61, 1980.
36. **Morris, N. P., Fessler, L. I., Weinstock, A., and Fessler, J. H.,** Procollagen assembly and secretion in embryonic chick bone, *J. Biol. Chem.,* 251, 7137, 1975.
37. **Fessler, L. I., Morris, N. P., and Fessler, J. H.,** Procollagen: biological scission of amino and carboxyl extension peptides, *Proc. Natl. Acad. Sci. U.S.A.,* 72, 4905, 1975.
38. **Davidson, J. M., McEneany, L. S. G., and Bornstein, P.,** Intermediates in the conversion of procollagen to collagen. Evidence for step-wise limited proteolysis of the COOH-terminal peptide extension, *Eur. J. Biochem.,* 81, 349, 1977.
39. **Burke, J. M., Balian, G., Ross, R., and Bornstein, P.,** Synthesis of type I and III procollagen and collagen by monkey aortic smooth muscle cells in vitro, *Biochemistry,* 16, 3243, 1977.
40. **Fessler, L. I. and Fessler, J. H.,** Characterization of type III procollagen from chick embryo blood vessels, *J. Biol. Chem.,* 254, 233, 1979.
41. **Smith, B. D., McKenney, K. H., and Lustberg, T. J.,** Characterization of collagen precursors found in rat skin and rat bone, *Biochemistry,* 16, 2980, 1977.
42. **Gould, B. S.,** Collagen biosynthesis, in *Treatise on Collagen,* Gould, B. S., Ed., Academic Press, New York, 1968, 139.
43. **Goldberg, B. and Green, H.,** Collagen synthesis on polyribosomes of cultured mammalian fibroblasts, *J. Mol. Biol.,* 26, 1, 1967.
44. **Boedtker, H., Frischauf, A. M., and Lehrach, H.,** Isolation and translation of calvaria procollagen messenger ribonucleic acids, *Biochemistry,* 15, 4765, 1976.
45. **Monson, J. M. and Goodman, H. M.,** Translation of chick calvarial procollagen messenger RNAs by a messenger RNA dependent reticulocyte lysate, *Biochemistry,* 17, 5122, 1978.
46. **Palmiter, R. D., Davidson, J. M., Gagnon, J., Rowe, D. W., and Bornstein, P.,** $NH_2$-terminal sequence of the chick proα1 (I) chain synthesized in the reticulocyte lysate system, *J. Biol. Chem.,* 254, 1433, 1979.
47. **Graves, P., Olsen, B. R., Fietzek, P. P., Prockop, D. J., and Monson, J. M.,** Comparison of the $NH_2$-terminal sequences of chick type I preprocollagen chains synthesized in an mRNA-dependent reticulocyte lysate, *Eur. J. Biochem.,* 118, 363, 1981.
48. **Olsen, B. R. and Berg, R. A.,** Post-translational processing and secretion of procollagen in fibroblasts, in *Secretory Mechanisms,* Hopkins, C. R., and Duncan, C. J., Eds., Society for Experimental Biology, Cambridge University Press, London, 1979, 57.
49. **Kivirikko, K. I. and Myllylä, R.,** The hydroxylation of prolyl and lysyl residues, in *The Enzymology of Posttranslational Modification of Proteins,* Friedman, R. B. and Hawkins, H. C., Eds., Academic Press, New York, 1980, 53.
50. **Adams, E. and Frank, L.,** Metabolism of proline and the hydroxyprolines, *Annu. Rev. Biochem.,* 49, 1005, 1980.
51. **Tuderman, L., Oikarinen, A., and Kivirikko, K. I.,** Tetramers and monomers of prolyl hydroxylase in isolated chick-embryo tendon cells, *Eur. J. Biochem.,* 66, 615, 1977.
52. **Stassen, F. L. H., Cardinale, G. J., McGee, J. O'D., and Udenfriend, S.,** Prolyl hydroxylase and an immunologically related protein in mammalian tissues, *Arch. Biochem. Biophys.,* 160, 340, 1974.
53. **Tuderman, L.,** Developmental changes of prolyl hydroxylase activity and protein in chick embryo, *Eur. J. Biochem.,* 66, 615, 1976.
54. **Chen-Kiang, S., Cardinale, G. J., and Udenfriend, S.,** Homology between a prolyl hydroxylase subunit and a tissue protein that crossreacts immunologically with the enzyme, *Proc. Natl. Acad. Sci. U.S.A.,* 74, 4420, 1977.
55. **Berg, R. A., Kao, W. W.-Y., and Kedersha, N. L.,** The assembly of tetrameric prolyl hydroxylase in tendon fibroblasts from newly synthesized α-subunits and from preformed cross-reacting protein, *Biochem. J.,* 189, 491, 1980.

56. **Pinnell, S. R., Krane, S. M., Kenzora, J. E., and Glimcher, M. J.,** A heritable disorder of connective tissue. Hydroxylysine-deficient collagen disease, *New Engl. J. Med.*, 286, 1013, 1972.
57. **Krane, S. M., Pinnell, S. R., and Erbe, R. W.,** Lysyl-protocollagen hydroxylase deficiency in fibroblasts from siblings with hydroxylysine-deficient collagen, *Proc. Natl. Acad. Sci. U.S.A.*, 60, 2899, 1972.
58. **Kivirikko, K. I. and Myllylä, L.,** Collagen glycosyltransferases, *Int. Rev. Connect. Tissue Res.*, 8, 23, 1979.
59. **Brownell, A. G. and Veis, A.,** Intracellular location of the glycosylation of hydroxylysine of collagen, *Biochem. Biophys. Res. Commun.*, 63, 341, 1975.
60. **Harwood, R., Grant, M. E., and Jackson, D. S.,** Studies on the glycosylation of hydroxylysine residues during collagen biosynthesis and the subcellular localization of collagen galactosyl transferase and collagen glucosyl transferase in tendon and cartilage cells, *Biochem. J.*, 152, 291, 1975.
61. **Clark, C. C. and Kefalides, N. A.,** Carbohydrate moieties of procollagen: incorporation of isotopically labeled mannose and glucosamine into propeptides of procollagen secreted by matrix-free chick embryo tendon cells, *Proc. Natl. Acad. Sci. U.S.A.*, 73, 34, 1976.
62. **Clark, C. C. and Kefalides, N. A.,** Localization and partial composition of the oligosaccharide units on the propeptide extensions of type I procollagen, *J. Biol. Chem.*, 253, 47, 1978.
63. **Murphy, W. H., von der Mark, K., McEneany, L. S. G., and Bornstein, P.,** Characterization of procollagen-derived peptides unique to the precursor molecule, *Biochemistry*, 14, 3243, 1975.
64. **Pesciotta, D. M., Dickson, L. A., Showalter, A. M., Eikenberry, E. F., de Crombrugghe, B., Fietzek, P. P., and Olsen, B. R.,** Primary stucture of the carbohydrate-containing regions of the carboxyl propeptides of type I procollagen, *FEBS Lett.*, 125, 170, 1981.
65. **Tanzer, M. L., Rowland, F. N., Murray, L. W., and Kaplan, J.,** Inhibitory effects of tunicamycin on procollagen biosynthesis and secretion, *Biochem. Biophys. Acta*, 500, 187, 1977.
66. **Duksin, D. and Bornstein, P.,** Impaired conversion of procollagen to collagen by fibroblasts and bone treated with tunicamycin an inhibitor of protein glycosylation, *J. Biol. Chem.*, 252, 955, 1977.
67. **Tabas, I., Schlesinger, S., and Kornfeld, S.,** Processing of high mannose oligosaccharides to form complex type oligosaccharides on the newly-synthesized polypeptides of the vesicular stomatitis virus G protein and the IgG heavy chain, *J. Biol. Chem.*, 253, 716, 1978.
68. **Lukens, L. N.,** Time of occurrence of disulfide linking between procollagen chains, *J. Biol. Chem.*, 251, 3530, 1976.
69. **Uitto, J., Hoffmann, H.-P., and Prockop, D. J.,** Role of triple-helical conformation in the secretion of procollagen: accumulation of non-helical peptides containing *cis*-hydroxyproline in the cisternae of the rough endoplasmic reticulum, *Science*, 190, 1202, 1975.
70. **Trelstad, R. L.,** Vacuoles in the embryonic chick corneal epithelium, an epithelium which produces collagen, *J. Cell Biol.*, 48, 689, 1971.
71. **Bruns, R. R., Hulmes, D. J. S., Therrien, S. F., and Gross, J.,** Procollagen segment-long-spacing crystallites: their role in collagen fibrillogenesis, *Proc. Natl. Acad. Sci. U.S.A.*, 76, 313, 1979.
72. **Prockop, D. J. and Guzman, N. A.,** Collagen diseases and the biosynthesis of collagen, *Hosp. Pract.*, 12, 61, 1977.

Chapter 2

# DETERMINATION OF COLLAGEN SYNTHESIS IN TISSUE AND CELL CULTURE SYSTEMS

**Beverly Peterkofsky, Mario Chojkier, and John Bateman**

## TABLE OF CONTENTS

# I. AREAS TO BE COVERED

This chapter covers all aspects of experiments designed to measure the rate of collagen synthesis in whole animals, isolated tissues, or cell cultures. In general,the discussion will concentrate on those methods developed or used in our laboratory, although other procedures are included.

Collagen synthesis can be measured by either of two general procedures. In one, radioactivity incorporated from precursor proline into hydroxyproline can be measured provided that proline hydroxylation is not impaired. In the other, the use of bacterial collagenase digestion to measure the collagen polypeptide eliminates the necessity for complete proline hydroxylation. With this method, however, it is essential that the collagenase preparation be free of nonspecific protease activity. If there is a significant amount of collagen degradation occurring during the time interval used to label tissues or cells with radioactive amino acid, then the incorporation into macromolecular collagen observed will be the rate of accumulation of the protein rather than the rate of synthesis. This will be true regardless of whether synthesis is expressed in relative or absolute terms. Therefore, we describe methods to determine the relative and absolute rates of collagen synthesis and the rate of degradation. The possibility that changes in the rate of transport of the amino acid precursor may lead to errors in measurement of these parameters is discussed. Finally, since hydroxylation of proline occurs during collagen synthesis and this reaction may alter the rates of accumulation and secretion of the protein,[1,2] we have included a description of a new method for determining the extent of proline hydroxylation concomitantly with measuring collagen synthesis.

# II. BIOCHEMICAL METHODS FOR MEASURING COLLAGEN SYNTHESIS

## A. Incorporation of Radioactivity from Proline into Hydroxyproline in Collagen

Until recently, the most widely used method to measure collagen synthesis exploited the fact that hydroxyproline is almost uniquely found in collagen, with only small amounts occurring in elastin and nonconnective tissue proteins, such as the Clq component of complement[3] and the enzyme acetylcholinesterase.[4]

### 1. Factors for Consideration

#### a. Dissociation of Proline Hydroxylation

The disadvantage of this method is that two reactions are measured: collagen synthesis and proline hydroxylation. It is possible to have dissociation of the two processes when factors required for the prolyl hydroxylase reaction, such as ascorbate or iron, become rate limiting. In that case, collagen polypeptide synthesis may take place, but the proline residues will not be hydroxylated. If hydroxyproline is used to measure the rate of collagen synthesis, the value will be underestimated. This occurred in the early studies of collagen synthesis in 3T6 cultured mouse cells[5] because the experiments were carried out in the absence of ascorbic acid and cultured mouse cells do not synthesize ascorbate.[6–8] Recent experiments have shown that stationary phase cultures of several mouse cell lines, such as 3T6 and L-929, can hydroxylate proline almost completely in the absence of ascorbate.[9,10] Such ascorbate-independent proline hydroxylation can take place because of the presence of an alternate reducing cofactor found in the microsomal membrane.[11,12] On the other hand, logarithmic-phase cells require ascorbate and have lower levels of the reducing cofactor.[10,11] These recent findings account for the fact that radioactive hydroxyproline could be measured only in stationary phase cells in Green and Goldberg's experiments,[5] an observation which led to the erroneous conclusion that only confluent or nongrowing cells synthesized collagen. In fact, logarithmic-phase cells of various lines have a higher absolute rate of collagen synthesis than stationary phase cells, while the relative rate does not change markedly during growth.[10] Therefore, much more efficient labeling of collagen can be achieved by using logarithmic-phase cells.

#### b. Ascorbate Requirement for Proline Hydroxylation in Cell Culture

In addition to the fact that most cultured cells do not synthesize ascorbate, the compound may become rate limiting for hydroxylation even if added to the medium because of its instability due to oxidation. The half-life of 0.25 m*M* ascorbate in Eagles minimal essential medium without (MEM-0) or with (MEM-10) 10% fetal calf serum incubated in culture dishes under 95% air/5% $CO_2$ was determined to be 6 and 9 hr, respectively (see Figure 1).[10] Ascorbate is oxidized initially, and by 3 hr the reduced form represents only 70% of the remaining material. This proportion remains fairly constant throughout the decay curve. The half-life is even shorter in Waymouth's medium in an atmosphere of 95% $O_2$/5% $CO_2$.[13] Ascorbate in culture medium also deteriorates when stored at 4 to 5°C in screw-capped bottles.[13,14] Ascorbate solutions stored at −20°C are stable only at relatively high concentrations such as 10 m*M*.[10] Analysis of hydroxyproline or radioactivity incorporated into this imino acid can be used as a valid method to measure collagen synthesis if precautions are taken to assure that an adequate concentration of ascorbate is present during the labeling period with radioactive proline. This will depend on the half-life for ascorbate in the particular culture medium being used and on the concentration of ascorbate required for proline hydroxylation.

In cultured chick embryo bone cells, the apparent $K_m$ for hydroxylation and collagen secretion, which is directly related to hydroxylation, was approximately 0.8 $\mu M$.[1] This value may vary with different systems, for in BALB 3T3 mouse cells the apparent $K_m$ determined by measuring collagen secretion was almost four times higher (3 $\mu M$), with saturation being achieved at 10 $\mu M$ (see Figure 2). The secretion which occurs in the absence of ascorbate is due to ascorbate-independent proline hydroxylation in untransformed cells, such as BALB 3T3, chick embryo fibroblasts,

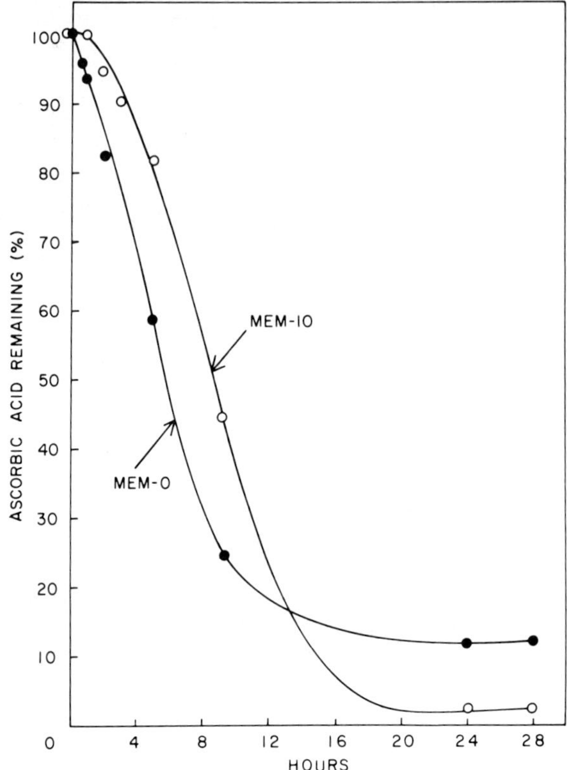


FIGURE 1. Rate of disappearance of ascorbate in tissue culture media. Ascorbate (0.25 m*M*) was incubated in 10 mℓ of MEM-0 (●——●) or MEM-10 (○——○) at 37°C under 95% air/5% $CO_2$. At the time intervals indicated in the figure, 0.5-mℓ samples were removed, an equal volume of 10% TCA was added, and the MEM-10 samples were centrifuged to remove the protein precipitates. Portions of the acid-soluble fractions were assayed for ascorbic acid by a method using reaction with dinitrophenylhydrazine. The curves represent the sum of ascorbate, and the oxidized products dehydroascorbate and diketogulonic acid. After 3 hr, the oxidized products represent 30% of the total. (From Peterkofsky, B., *Arch. Biochem. Biophys.*, 152, 318, 1972)

human diploid fibroblasts, and guinea pig fibroblasts at about 10 to 20% of the maximum hydroxylation obtained in the presence of saturating ascorbate.[1,7,15] As mentioned above, 3T6 and L-929 cells are almost completely ascorbate-independent for hydroxylation in stationary phase, while in logarithmic phase they behave like untransformed cells.[9–11] Still other cell lines, such as Kirsten sarcoma virus transformed BALB 3T3 (Ki-3T3) cells, exhibit a high degree of ascorbate-independent proline hydroxylation throughout growth.[7]

*c. Cytotoxicity of Ascorbate*

Before leaving the subject of ascorbic acid, it might be pointed out that this compound generally is not a growth requirement for cultured cells.[10] In vivo ascorbic acid is a vitamin for those animals incapable of its synthesis because it is required for skeletal growth and maintenance, in addition to other vital functions. Exceptions

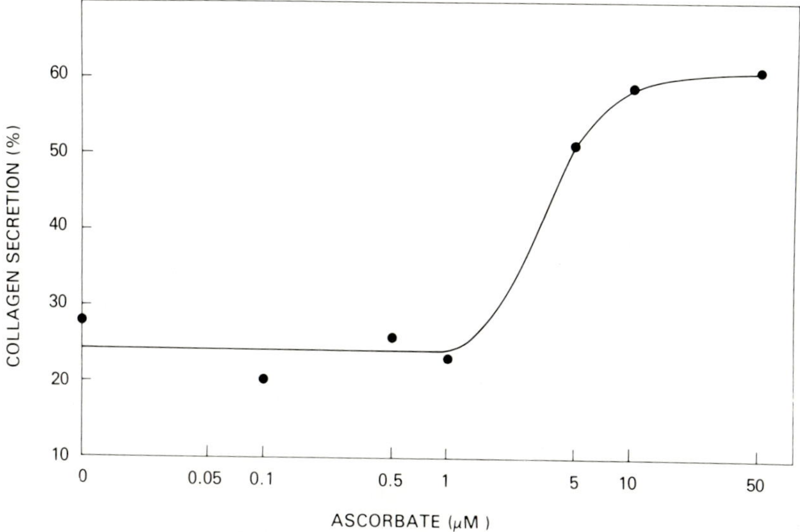


FIGURE 2. Collagen secretion by BALB 3T3 cells as a function of ascorbate concentration. BALB 3T3-P3 (subclone of 3T3A31) cell cultures were initiated on day 0 at 5 $\times$ $10^5$ cells per 100 $\times$ 20 mm dish in 15 mℓ of MEM-10. On day 2, when the density was 3 $\times$ $10^6$ cells per dish, growth medium was removed, and the cell layers were rinsed twice with MEM-*0* and incubated at 37°C under 95% air/5% $CO_2$ with 3.0 mℓ of MEM-*0* containing varying concentrations of sodium ascorbate, as indicated in the figure. *L*-[$^{14}$C]Proline (1.5 μCi, 0.3 μmol) was added to each dish and incubation was continued for 2.5 hr. The cells and medium were collected and separately processed, and collagen was measured by collagenase digestion as described in Section II.B. The percentage secreted was calculated from the fraction of radioactive collagen in the medium compared to the total collagen in cells plus medium.

to these observations are one report which suggests that a low concentration of ascorbate (28 μ*M*) stimulates the growth of human fibroblasts approximately two-fold[16] and another which indicates that ascorbate addition to cultures of chick embryo chondrocytes increased the number of cells without increasing the amount of protein per culture.[17] In fact, when cells are plated at low cell densities (4 $\times$ $10^4$/cm$^2$), ascorbate is cytotoxic.[16,18] Since ascorbate transport is quite rapid,[1] the compound need not be added until 15 to 30 min before the labeling period, thus allowing cells to reach densities resistant to the toxic effect of ascorbate at the concentrations necessary to obtain hydroxylation.

### 2. *Experimental Procedures*

The procedures for labeling proteins of tissues or cells with radioactive amino acids will be described further on under the section dealing with collagen measurement by means of bacterial collagenase digestion. The major addition to that protocol is the necessity of adding ascorbate just prior to the labeling period. The radioactively labeled protein must first be hydrolyzed in order to measure hydroxyproline. Tissue or cell homogenates are dialyzed extensively or the proteins in them are precipitated with trichoroacetic acid (TCA) in order to remove most of the unincorporated radioactive proline. The proteins are hydrolyzed for 16 hr at 100°C or for 3 hr under pressure (15 lb, 120°C) in an autoclave,[19] and the HCl is removed under vacuum on a rotary evaporator or a Buchler Evapomix apparatus. The hydrolyzate can then be analyzed for radioactive hydroxyproline by ion exchange chromatography on a Dowex-50 column using 1 *N* HCl for elution,[19,20] an amino acid analyzer with a stream-splitting device,[21] high-voltage electrophoresis,[22] thin layer chromatography,[23] or by a technique using Chloramine-T oxidation and toluene extraction of pyrrole which is formed from the oxidized product of hydroxyproline after heating.[24,25]

### B. Digestion of Radioactive Collagen by Purified Bacterial Collagenase

#### *1. Properties of the Enzyme*

The use of bacterial collagenase allows collagen polypeptide synthesis to be measured independent of the process of proline hydroxylation. There have been several reviews describing the specificity and properties of the enzyme from *Clostridium histolyticum*[26,27] so we will not discuss this aspect in great detail. The enzyme will cleave peptide bonds in the general sequence Pro-X-Gly-Pro-Y between X and Gly when X is proline, hydroxyproline, alanine, or a number of other amino acids. Its specificity is due to the repetitive occurrence of such sequences in collagen and their virtual absence in other proteins. The flexibility of cleavage with respect to the amino acid in the X position explains the fact that the enzyme can cleave either hydroxylated or unhydroxylated collagen. Preparations of the enzyme appear to contain several different collagenases, some of which preferentially cleave native collagen and some which cleave the denatured protein or synthetic peptides as well.[25–27]

The most important aspect of using this technique for measuring collagen is that the collagenase preparation should be free of any nonspecific proteolytic activity or that such activity, if present in small amounts, should be suppressible by an inhibitor which will not affect collagenase activity. Many commercial preparations of purified collagenase have been found to contain a substantial amount of noncollagen proteolytic enzyme activity.[25,28–30] A more detailed discussion and evaluation of the extent of contamination of several commercial preparations of bacterial collagenases will be presented elsewhere.[31]

The most abundant contaminant is clostripain, previously called amidase/esterase, an enzyme with a trypsin-like activity which can be inhibited by SH reagents.[32] Not all proteolytic activity in a commercial preparation could be inhibited by *N*-ethyl maleimide (NEM), indicating that there were still other proteases present besides clostripain.[25] Gel filtration of the preparation removed all of the nonsuppressible proteases and all but a trace of clostripain, but this activity could be completely inhibited by NEM.[25] By including NEM in the reaction mixture for digestion of radioactive collagen, absolute specificity for collagen was obtained.

Protease activity directed against noncollagenous proteins can be detected by incubating radioactive tryptophan-labeled proteins with the collagenase preparation under exactly the same conditions as used in the assay to measure collagen. The basis for this assay is the fact that the portion of collagen which forms the triple helix and is the substrate for bacterial collagenase does not contain tryptophan. The presence of radioactive tryptophan in the acid-soluble fraction after the incubation is indicative of a contaminating protease.

Besides its specificity for the unique sequence of collagen, collagenase possesses other characteristics which make it well suited to be used as an analytical reagent. It is extremely stable in the presence of calcium ion, allowing purification to be carried out at room temperature with 100% recovery of activity.[25] In our laboratory, fractions from the gel filtration column containing collagenase activity are pooled, divided into several portions, and lyophilized. The lyophilized material may be stored desiccated at 4°C for as long as 5 years with no loss of activity. Solutions

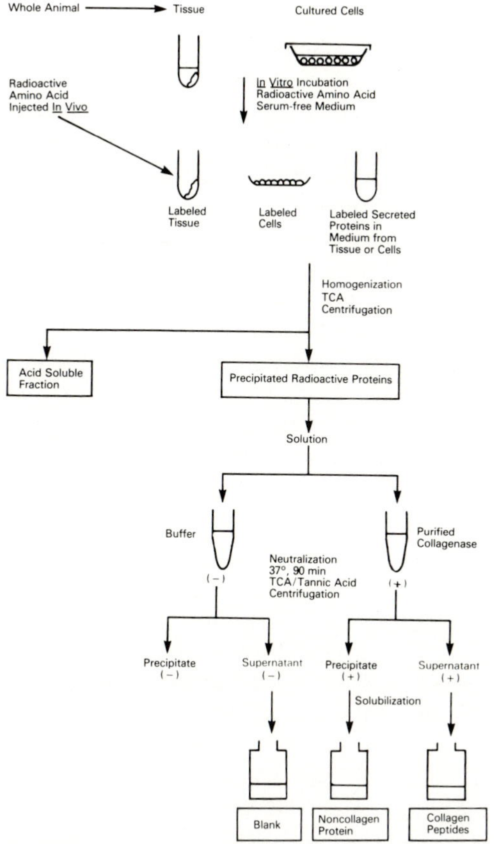


FIGURE 3. Schematic representation of the steps involved in measuring collagen synthesis in different biological systems. Details of each step are described in Section II.B.

reconstituted from the lyophilized preparation are then divided into small portions and stored at −20°C. Such solutions are stable for as long as 6 months provided they are thawed and refrozen only a few times.

### 2. *Experimental Procedure*

#### ***a. Outline of Procedure and Reagents Required***

Details of the procedure will be described as carried out currently in our laboratory and are outlined in Figure 3. This description includes several modifications of the collagenase digestion procedure which have been introduced by us since the method was originally described.[25]

In brief, a tissue or cell sample which has been labeled with a radioactive amino acid, usually proline, during the linear portion of a time vs. incorporation plot, is dispersed in some way. Unincorporated radioactive amino acid is separated from proteins by TCA treatment, and the TCA precipitated protein is redissolved in NaOH. Equal portions of this solution are neutralized and incubated without or with collagenase, and the proteins are reprecipitated with a mixture of TCA and tannic acid. The peptides derived from collagen remain in the TCA/tannic acid-soluble fraction, while noncollagen proteins are in the reprecipitated fraction of the sample which had collagenase added to it. The minus enzyme control is used to correct for any free radioactive amino acid which had been bound to proteins and released by dissolution in NaOH. A modification of this procedure has been introduced in which the final TCA/tannic acid precipitates are collected on nitrocellulose filters and the filters counted.[33,34] The amount of collagen is calculated from the difference in radioactivity on the (−) and (+) filters. While this may be feasible in systems where the radioactivity incorporated into collagen is more than 25% of the incorporation into total protein, it is very inaccurate for systems such as cell cultures where only 10% or less of the incorporated radioactivity may be in collagen. Reagents — Only those reagents not commonly used are listed:

1. Tris® buffered saline — A solution of 0.11 *M* NaCl containing 0.05 *M* Tris-HCl and adjusted to pH 7.4
2. HEPES buffer, 1 *M*, pH 7.2 — Dissolve 23.83 g HEPES (Calbiochem®) in 80 mℓ of $H_2O$, gradually add 10 *N* NaOH to bring the pH to 7.2, and adjust the volume to 100 mℓ with $H_2O$ — Store at −20°C
3. 10% TCA/0.5% tannic acid — Store at 4°C for no longer than 1 week — Tannic acid tends to become oxidized and acquire an intense yellow color after longer periods
4. NaOH, 0.2 *N* — Dilute a standardized solution of 1.00 *N* NaOH (Fisher®, stored at −20°C) — Store in small portions at −20°C
5. HCl, 0.15 *N* — Dilute a standardized solution of 1.00 *N* HCl (Fisher®, stored at 4°C). Store in small portions at −20°
6. NEM, 62.5 m*M* — Adjust the pH to about 6 by adding small amounts of solid $NaHCO_3$ and mixing well — Store in small portions at −20°C
7. Elution buffer — A solution of 0.05 *M* tris-HCl, pH 7.6 containing 5 m*M* $CaCl_2$
8. Purified collagenase — An enzyme preparation containing 600 to 800 μg protein per milliliter in elution buffer, purified from Worthington® CLSPA collagenase as described previously[25] — Protein concentration is measured with BioRad® reagent using immunoglobin as a standard
9. Chick embryo carrier protein — Ten-day-old embryos are removed from eggs and placed in 0.15 *M* NaCl (0°C) — They are rinsed with saline, blotted dry, weighed, minced, and homogenized with one volume of $H_2O$ — The homogenate is centrifuged at 43,000 × g for 30 min and the supernatant solution removed and dialyzed against 20 volumes of $H_2O$ at 4°C — The retentate is incubated at 37°C for 1 hr to allow inactivation of proteolytic enzymes and then centrifuged at 43,000 × g for 10 min — The protein concentration of the supernatant solution is determined and portions distributed into tubes and lyophilized — The residue is dissolved in 0.02 *M* Tris-HCl, pH 7.6, to give a protein concentration of 10 mg/mℓ

***b. Labeling of Tissues and Cell Cultures with Radioactive Amino Acids***

Intact chick embryos are frequently used as experimental animals because of their rapid rate of collagen synthesis. They may be labeled by injection into a chorioallantoic artery which results in rapid incorporation, but requires extreme care.[19] A simpler procedure is to make a small hole in the shell above the air space and to place a small volume (up to 0.2 mℓ) of a solution containing 5 to 20 μCi of radioactive proline on top of the shell membrane. The hole is covered with cellophane tape, and the egg is reincubated in an upright position with the air space at the top. The position of the air space can be determined using an egg candler. The rate of [$^{14}$C] proline uptake into embryos was found to vary with age so that the labeling period must be adjusted in order to obtain a linear time course of incorporation.[35]

Intact tissues, such as chick embryo frontal or calvarial bones,[36,37] tendons,[38] lens,[39] or aorta,[40] have been used in short-term in vitro culture systems. Fusion of chick embryo frontal and parietal bones to form the calvarium occurs at about day 16.[41] We have used Eagles MEM-0 for chick embryo frontal bone cultures. With bones from 14- to 16-day-old embryos, linear incorporation of radioactive proline proceeds for approximately 2 to 3 hr. Adequate incorporation is obtained with one to two bones in 0.5-mℓ medium containing 1 μCi of [$^{14}$C]proline. The amount of radioactive proline required for the dual-labeled proline method to measure collagen synthesis and proline hydroxylation concommitantly is discussed in Section IV. The percent collagen synthesis in bones from different-aged embryos obtained under these conditions varies exactly as it does in ovo, with maximum synthesis occurring after 14 days.[42] The percentage of collagen observed in vitro, however, was lower than it was in the same tissue in the intact embryo.[35]

A variety of conditions have been used to label cultured cells with radioactive amino acids. It is often thought that using the highest specific radioactivity possible will result in the greatest amount of labeling; this is not always the case. The amount of radioactivity from proline incorporated into protein is dependent not only on the rate of protein synthesis, but on the rate of proline transport. Since the $K_m$ for transport of proline, 6.5 m*M*, as well as for other amino acids is relatively high,[43,44] the rate of uptake is quite low at amino acid concentrations resulting from addition of isotope alone. For example, if 0.5 to 5.0 μCi of proline with a specific activity of 200 μCi/μmol were used per milliliter of serum-free medium, the final concentration of proline in the medium would be approximately 0.0025 to 0.025 m*M*. If serum were present, the concentration would be higher, since serum contains about 0.15 m*M* proline.[45] At proline concentrations below the $K_m$ value, the rate of incorporation into protein would be linear for only a short time because as the proline concentration decreased, so would the rate of transport. This situation is suitable for very short-term pulse labeling, but if it is desirable to achieve labeling over a longer time interval, it is advantageous to dilute the radioactive proline with unlabeled proline and thus maintain a higher rate of transport. In Figure 4, an experiment is illustrated in which the amount of radioactive proline added to a culture was kept constant, but the total concentration of proline was increased. [$^{14}$C]Proline transported into the acid soluble pool in the absence of protein synthesis was measured. If protein synthesis were not inhibited, the amount of uptake would be equal to the total radioactivity in the acid-soluble and insoluble fractions. It may be seen in Figure 4 that when radioactive proline in the medium was diluted up to 200-fold, there was no decrease in the amount of radioactivity observed in the cells. This resulted from the rate of transport being increased in a directly inverse relationship to the dilution over the range 0.025 to 0.5 m*M* proline. Eventually, further dilution resulted in decreased intra-cellular radioactivity, but even at 4 m*M*, which is a 1600-fold dilution of the isotope, there was only a 36% decrease in the apparent uptake. The main point to be derived from this experiment is that diluting a high specific activity isotope does not necessarily lead to a reduction in incorporation of radioactivity into cells. The amount and specific activity of the amino acid precursor thus should be altered according to the protocol of each particular experiment.

Before adding the radioactive amino acid to cultures, growth medium is removed

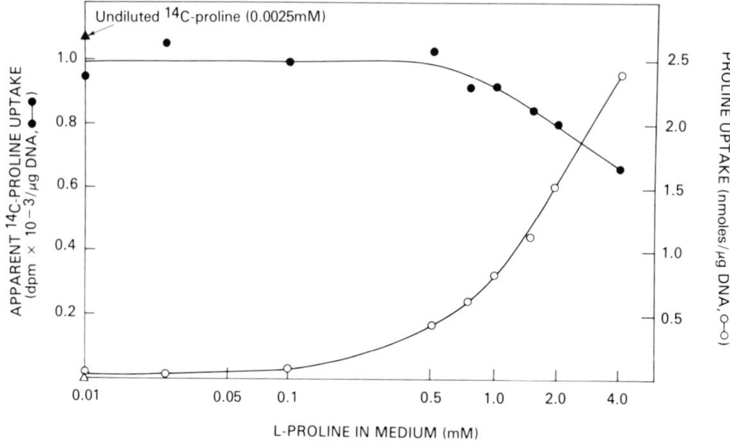


FIGURE 4. The effect of varying exogenous proline on *L*-[$^{14}$C]proline transport into Ki-3T3 cells. Cultures were initiated at $8 \times 10^5$ cells per $100 \times 20$ mm dish in MEM-10 and grown at 37°C under 95% air/5% $CO_2$. When cells reached logarithmic phase ($4.2 \times 10^6$ cells per dish), growth medium was removed, the cell layers were rinsed with MEM-0 (37°C) twice, and 3.0 mℓ of MEM-0 containing 0.1 m*M* ascorbate and 0.1 m*M* cycloheximide was added. The cultures were preincubated for 15 min at 37°C as above, and then 0.15 mℓ of solutions containing 10 μCi/mℓ of *L*-[$^{14}$C]proline at varying concentrations of proline were added to replicate cultures, and incubation was continued for 15 min. Uptake was stopped by immediately removing the medium and rinsing the dishes twice with tris-buffered 0.11 *M* NaCl at 0°C. The cell layer was scraped off in 1.5 mℓ 0.05 *M* Tris-HCl, pH 7.6, with a rubber policeman. The dish was rinsed with an additional 1.5 mℓ of Tris-buffer, and the rinse was combined with the cell lysate. The lysate was sonicated and TCA added to give a final concentration of 5%. After centrifugation of the precipitated protein, resuspension of 5% TCA, and recentrifugation, radioactivity in the TCA-soluble fraction was measured by liquid scintillation spectrometry. Radioactivity was corrected for the amount of DNA in each dish and is presented in the figure as apparent [$^{14}$C]proline uptake (•——•). To determine the actual proline uptake as nanomoles of proline per microgram DNA, the radioactivity was divided by the specific activity of [$^{14}$C] proline at each different concentration (○——○).

from logarithmic phase cells and the cell layer is rinsed with MEM-0 (37°C) by carefully pipetting the solution against the side of the dish. The dish is gently swirled and the medium siphoned off. This process is repeated once more and then 3.0 mℓ of MEM-0 is added. Labeling cell cultures in serum-free medium avoids the problem of having a large concentration of serum protein to contend with in processing the medium and also eliminates a component of serum which interferes with precipitation of TCA insoluble material.[46] If ascorbate is to be added, the cultures are incubated at 37°C under 95% air/5% $CO_2$ for 15 min to allow uptake by the cells. Radioactive proline is added and incubation at 37°C resumed for the desired time interval.

***c. Preparation of Samples for Digestion by Collagenase***

Once the incubation of tissues or cells is stopped, all procedures should be carried out at 0 to 4°C. In the case of tissues labeled in vivo or in vitro, these must be homogenized. For soft tissues such as liver, this can be accomplished with glass tissue grinders such as a Dounce homogenizer, but for bone or cartilage, a more vigorous means must be used. A Polytron® homogenizer (Brinkmann®) set at maximum speed is quite efficient for hard tissues. For in vitro cultures the tissue may be homogenized directly in the medium or, preferably, the medium may be removed and treated separately, especially if secretion is to be measured.[36] The procedure for treating the medium will be described following the procedure for tissue processing.

If the medium is removed, the tissue is rinsed twice with buffered saline containing 1 m*M* *L*-proline, to aid in removing unincorporated radioactive proline, and then minced with a scissors. With the Polytron® generator PT 10 ST, a minimum volume of 1.5 mℓ of added fluid must be used to homogenize the tissue in a 15 × 125 mm test tube; we use 0.05 *M* Tris-HCl, pH 7.6, containing 1 m*M* proline. Bone tissue tends to become trapped within the generator, but can be removed by running the Polytron® with 1.5 mℓ of buffer for exactly the same time interval and speed as the initial homogenization. This rinse is added to the homogenate. One fourth volume of 50% TCA is added to the homogenate to give a final concentration of 10%, and the suspension is kept at 0°C for 5 min. After the precipitate is collected by centrifugation at 1000 × g for 5 min, it is resuspended with 1.0 mℓ of 5% TCA containing 1 m*M* proline and recentrifuged. The supernatant fraction is removed, and this procedure repeated at least twice more. The precipitate is then dissolved with 0.2 *N* NaOH, usually 0.6 mℓ. For bone tissue, it may be necessary to heat the suspension for 5 min at 100°C in order to get complete solubilization of the proteins.

For processing the medium, which contains almost no protein, a carrier protein preparation which can be reprecipitated by TCA/tannic acid must be added. Bovine serum albumin is not precipitable once it has been dissolved in NaOH, so we use a preparation of soluble proteins from 10-day-old chick embryo of which 1.5 mg is added to each medium sample just prior to the addition of TCA (final concentration of 10%). From that point on, the sample is treated in the same manner as the tissue sample.

The steps followed for cell cultures are similar to those described above. At completion of the labeling period, cultures are removed from the incubator and quickly chilled to 0°C on ice. If collagen secretion is not being measured, cells are scraped from the dish directly in the medium using a rubber policeman. Otherwise, the culture is separated into cell and medium fractions. The cell layer is rinsed twice with 2 mℓ per dish of Tris® buffered saline containing 1 m*M* proline and these rinses are added to the medium. The cells are scraped off the dish in 1 mℓ buffered saline using a rubber policeman, and the suspension is transferred to a conical tube. this tube should be plastic, since glass may shatter in the subsequent sonication step. The dishes are rinsed with a small volume of buffered saline, and the solution is added to the cell suspension which is then centrifuged at 1000 × g for 5 min. The supernatant solution is added to the medium, and 0.5 mℓ of 0.05 *M* Tris-HCl, pH 7.6/1 m*M* proline, is added to the cell pellet. If the sample is to be stored at −20°C at this point, it should remain unmixed. Otherwise, resuspend the cells and sonicate for 25 to 30 sec at 25% of maximum voltage with the needle probe of the sonicator, keeping the sample immersed in an ice bath. In our original procedure, ribonuclease was added to release labeled amino acid from aminoacyl-tRNA, but this step has been dispensed with since cell cultures were found not to contain detectable amounts of prolyl-tRNA after TCA precipitation.[43] Add an equal volume of 20% TCA/2 m*M* proline to the sonicate, mix well, let stand at 0°C for 5 min, and then centrifuge at 1000 × g for 5 min. Thereafter the procedure is exactly as described above for tissue samples, except that it is not necessary to heat after resuspending the precipitate in 0.2 *N* NaOH. The precipitate derived from 4 to 10 × $10^6$ cells will usually dissolve completely at room temperature.

The medium from cell cultures is treated similarly to that of the tissue cultures.

Since several cultures are usually pooled, the volume of the medium may be quite large so that an initial step of dialysis and lyophilization is used to remove the bulk of the unincorporated radioactivity and to reduce the volume.[10] The residue is dissolved in a small volume of water and then treated in the same manner as the medium from tissue culture.

***d. Conditions for Digestion***

Two portions, usually 0.2 mℓ, of the 0.2 *N* NaOH solutions are transferred to 2-mℓ glass conical centrifuge tubes. The NaOH is neutralized by adding 0.1 mℓ of 1 *M* HEPES buffer, pH 7.2, plus sufficient 0.15 *N* HCl to adjust the pH to 6.8 to 7.4. The pH of a representative sample of a series should be measured. Since a large excess of collagenase is added and the reaction goes to completion rapidly, the pH need not be precisely at the optimum (7.2). With 1.5 mg of carrier chick embryo protein or the TCA precipitate from 4 to 10 × $10^6$ cells, 0.16 mℓ of 0.15 *N* HCl will bring the solution to the correct pH range. If a large amount of protein is present in the sample, however, there will be a smaller excess of NaOH and consequently less HCl will be needed.

After the solutions have been neutralized, they are chilled at 0°C. Reagent solutions are prepared without (−) or with (+) purified collagenase in the following proportions:

| | | (−) | (+) | Final conc in reaction mixture |
|---|---|---|---|---|
| $CaCl_2$ | 25 m*M* | 1 | 1 | 0.5 m*M* |
| NEM | 62.5 M*M* | 2 | 2 | 2.5 m*M* |
| Elution buffer | | 1 | 0 | |
| Purified collagenase | 600–800 μg/mℓ | 0 | 1 | 12–16 μg/mℓ |

A total of 40 μℓ of the "minus" solution is added to one of the duplicate sample solutions, and 40 μℓ of the "plus" solution is added to the other to give the final concentrations listed. The contents are mixed, and the tubes are incubated at 37°C for 90 min, after which they are chilled to 0°C, and 0.5 mℓ of 10% TCA/0.5% tannic acid is added to each tube. It was found necessary to add tannic acid to the TCA in order to achieve complete reprecipitation of undegraded proteins. After centrifugation at 1000 × g for 5 min, the soluble fraction is transferred to a counting vial. The precipitate is resuspended in 0.5 mℓ of 5% TCA/0.25% tannic acid, recentrifuged as above, and the soluble fraction added to the vial. Scintillation fluid suitable for aqueous samples is added, and the radioactivity is measured in a liquid scintillation spectrometer. To convert cpm to dpm, radioactive standards are prepared by adding a small volume of a calibrated [$^{14}$C] or [$^{3}$H]toluene solution to 1.5 mℓ of 5% TCA/0.25% tannic acid plus scintillation fluid.

The precipitate from the sample which had collagenase added to it is resuspended with 1 mℓ 6 *N* HCl, transferred to a tube with a Teflon®-lined screw cap, and hydrolyzed as described in the section on hydroxyproline measurement. A portion of the hydrolyzate is transferred to a counting vial, and scintillation fluid is added. In order to convert to dpm, radioactive standards are prepared with the same amount of 6 *N* HCl and scintillation fluid as used for the samples. Other tissue solubilizers may be used to dissolve the noncollagen protein precipitate, but if denaturation is not complete, total radioactivity will not be accurately measured because of self-absorption. This problem is most severe in the case of tritium-labeled proteins, but acid hydrolysis assures complete solubilization. DNA or protein assays can be carried out on the remaining NaOH solutions of the samples and incorporation may be expressed on the basis of either of these parameters.

*e. Calculation of Relative Rate of Collagen Synthesis*

1. Calculate the difference between the cpm obtained from the TCA/tannic acid soluble fractions of the "plus" and "minus" collagenase samples to obtain the amount of radioactivity in collagen (cpm)
2. Calculate the total dpm in collagen by correcting for the original volume and counting efficiency
3. Calculate the total dpm in noncollagen protein by correcting the cpm (minus background) in the HCl hydrolyzate for the total volume and counting efficiency. This value must be multiplied by 5.4 to correct for the fact that collagen has approximately 22.2% imino acids compared to an average of 4.1% in noncollagen proteins[35]
4. Calculate relative rate or percentage of collagen compared to total proteins by the formula:

$$\frac{\text{dpm in collagen}}{(\text{dpm in noncollagen protein} \times 5.4) + \text{dpm in collagen}} \times 100$$

It should be pointed out that most of the amino acids in the procollagen extension peptides are not in the characteristic sequence which is found in the region which forms the triple helix in the collagen molecule.[47] The exception is a short region of 30 to 40 amino acid residues in the N terminal propeptides. Collagenase digestion of pro$\alpha$-chains leaves the C terminal propeptide which has a molecular weight of about 30,000 and is TCA precipitable, at least in the unreduced form,[48] and two peptides, Col 1 and 2, from the *N*-propeptides of pro$\alpha_1$ (I) and (II). The helical region in the unreduced *N*-propeptide of procollagen III is resistant to collagenase, and the entire triple stranded 57,000-dalton *N*-propeptide will remain.[49] Digestion of chick and sheep pro$\alpha_2$ yields only Col 2.[50,51] The Col 1 peptides from the various pro$\alpha_1$-chains have molecular weight in the range of 10,000 to 11,000 and are probably precipitated by TCA. The Col 2 peptides, which span the C terminal region of the *N*-propeptide extension and the telopeptide end of the processed $\alpha$-chains, have molecular weights of about 3000 to 6000, and it has not been determined whether they are TCA precipitable. Those peptides which are TCA precipitable would be assayed as noncollagen protein, leading to an underestimation of the relative rate of collagen synthesis. Obviously, this problem does not exist in systems such as intact tissues where procollagen is rapidly converted to collagen,[38] and the propeptides are removed into the circulation (in vivo) or into medium (in vitro). In many cell cultures, however, procollagen is processed quite slowly, so that various intermediates and very little mature $\alpha$-chains are present, especially in short-term labeling experiments.[52]

Several groups have used correction factors to determine the percentage of procollagen synthesized in cell-free translation systems or organ and cell cultures.[42,53,54] The assumptions made are that the helical regions of the *N*-propeptides (Col 3) are digested by collagenase under the conditions of the assay and that Col 2 is precipitable by TCA. The possibility that Col 3 may not be digested was discussed above, and the ramifications of this are fully discussed by Breul and co-workers.[54] Although Col 1 is not present in sheep and chick pro$\alpha_2$, it may be present in other species,[47] so its contribution was not excluded. The formula derived here allows an approximation of the percentage of procollagen synthesized or accumulated and is similar to that of Breul and co-workers.[54] It is based on average values for the amino acid compositions of the propeptides from Types I and III procollagens.[21,48,49,55,56]

There are approximately 30 proline residues out of a total of approximately 443 amino acids in the propeptide extensions of a pro$\alpha$-chain. The amino acid content of procollagen is thus 17.5% rather than 22.2% as in collagen. Therefore, the noncollagen protein dpm must be corrected by 4.3 rather than 5.4 as in the formula given above. The proline residues of the propeptides represent 13.2% of the total proline residues in the major collagenase-sensitive region of the chain so the amount of radioactivity in the propeptides can be estimated by multiplying the collagen dpm by 0.13. This amount must be subtracted from the noncollagen protein before correcting by 4.3 and added to the collgen dpm to obtain procollagen dpm. The formula given above for calculating the relative rate of collagen is modified to include these corrections and allow an approximation of the percentage of procollagen synthesized or accumulated as follows:

$$\text{Percentage of collagen} = \frac{1.13\ (\text{collagen dpm}) \times 100}{4.3\ (\text{Noncollagen protein dpm} - 0.13 \times \text{collagen dpm}) + 1.13\ (\text{collagen dpm})}$$

## III. DETERMINATION OF ABSOLUTE RATE OF SYNTHESIS

### A. Use of the Specific Activity of the Amino Acid Precursor Pool to Calculate Absolute Rate of Synthesis

In addition to the effect of substrate concentration on proline transport discussed in Section II.B.2.b, differences in the rate of transport may occur when testing the effect of various drugs or hormones, comparing synthesis in animals at different ages or in cells at different stages of growth. For example, the transport of proline is greater in logarithmic- than in stationary-phase cells, greater in transformed than normal cells, and influenced by agents such as dbcAMP.[43] If the rate of transport is increased or decreased and the intracellular pool size remains constant, the specific radioactivity of the intracellular pool will be altered accordingly. Consequently the amount of radioactivity in the protein will be altered without any actual change in the rate of protein synthesis. While expressing synthesis as a relative rate, i.e., the rate of synthesis of a single protein compared to synthesis of all other cellular proteins, eliminates this problem, it is only a relative comparison of one experimental condition to another and often the absolute rate of synthesis must be determined.

In order to determine the absolute rate of synthesis of a protein in molar terms, the specific activity of the radioactive amino acid in the precursor amino acid pool of the cell or tissue must be measured. Ideally, the specific activity of the amino acid attached to tRNA should be used, but this pool is often too small to be measured (see Section II.B.2.c). There has been some controversy in the literature over whether the pool of free amino acids isolated from cells or tissues serves as the reservoir for protein synthesis. It is clear that at least part of this pool arises from degradation of proteins, as evidenced by the fact that free hydroxyproline can be found in it.[19] The remainder is derived from the biosynthesis of nonessential amino acids such as proline and from uptake of amino acids from the blood, in vivo, or from the culture medium, in vitro.

Several reports have presented data showing that the specific activity of an amino acid attached to tRNA was greater than that of the amino acid in the free amino acid pool in HeLa cell cultures or in rat liver in vivo.[57,58] Analysis of liver from newt labeled in vivo with leucine showed that the specific activity of nascent polypeptides attached to tRNA was eight- to tenfold higher than the specific activity of leucine in the free pool.[59] Another study which examined the role of the free pool in protein synthesis in HTC cells concluded that there is a single intracellular amino acid pool with amino acids entering from the extracellular medium and from protein degradation. This pool would be used as precursor for protein synthesis, but, in addition, it was postulated that amino acids could also be attached to tRNA without going through the pool and then become incorporated into protein.[60]

In direct opposition to these studies, there are numerous reports which strongly support the concept that the intracellular amino acid pool is the direct source of amino acids used to synthesize proteins. Loftfield and Harris induced ferritin synthesis in rats so that the amount in the liver remained constant for several weeks. A mixture of [$^{14}$C] labeled valine, isoleucine, and leucine was administered either by continuous infusion or by repeated injections for 3 days so that the amino acids attained constant specific activity.[61] Ferritin was isolated at several time intervals, and the specific activity of the three amino acids in the protein as well as in the free pool was determined. At all time points over the 3-day period, the specific activity of each amino acid in the protein was almost identical to the specific acitivity of that amino acid in the free pool. Fern and Garlick infused radioactive glycine in rats for 2 and 6 hr and determined the specific activity of glycine, as well as its metabolic product serine, in both the acid-soluble fraction of plasma and various tissues and in tissue proteins.[62] The ratio of serine/glycine specific activities was used to determine whether a common amino acid pool was used for protein synthesis. For each tissue, this ratio was almost identifical in the free pool and in protein. These results provide very persuasive evidence that the free pool provides the precursors for protein synthesis.

Prockop and co-workers carried out experiments in which [$^{14}$C]proline was injected into chick embryos, and the absolute rate of collagen synthesis was determined.[19] An average specific activity of free [$^{14}$C]proline during an interval of linear incorporation was used to determine the absolute rate of incorporation into collagen hydroxyproline. This rate was extrapolated to 3.6 μmol/day per embryo which was very similar to the rate of 4.2 μmol per embryo per day calculated from data of a previous study which had measured daily hydroxyproline accumulation in whole chick embryos. Thus, the use of the specific activity of proline in the free pool is a reasonable procedure to follow, since it gave a value close to that which could be measured by net accumulation of the product.

In an experiment carried out recently in our laboratory using transformed 3T3 cells, protein synthesis, as measured by [$^{14}$C]proline incorporation into the TCA insoluble fraction, was inhibited by addition of cycloheximide. It was found (see Table 1) that the total amount of radioactivity taken up by the cells was fairly constant in the absence or presence of cycloheximide, but that when protein synthesis was inhibited, then the amount in the free pool increased by approximately the amount that was not incorporated into protein. This experiment suggests that, at least in cultured cells, there is direct precursor-product relationship between the free pool and protein. This is not to imply that there are not complexities in using this approach, especially in whole animal or intact tissue systems. In such systems the existence of another compartment, the extracellular fluid, further complicates the transport process and the validity of using the free pool specific activity depends

**Table 1**
**ACCUMULATION OF [$^{14}$C]PROLINE IN THE ACID-SOLUBLE POOL AS A RESULT OF INHIBITION OF PROTEIN SYNTHESIS BY CYCLOHEXIMIDE**

| | [$^{14}$C]Proline incorporated (dpm × $10^{-3}$/30 min) | | |
|---|---|---|---|
| Cycloheximide | Acid soluble | Protein | Total |
| 0 | 101.1 | 47.0 | 148 |
| 0.05 | 146.1 | 1.6 | 148 |
| 0.10 | 161.1 | 1.1 | 162 |
| 0.30 | 152.1 | 0.5 | 153 |

*Note:* Ki-3T3 cells in logarithmic phase were incubated for 30 min with 1.5 μCi *L*-[$^{14}$C]proline (20 μCi/μmol) under conditions similar to those described in the legend to Figure 2, at 0.1 m*M* ascorbate. Radioactivity in the acid soluble and insoluble (protein) fractions was measured. Results are the average of duplicate samples, each consisting of two dishes (100 × 20 mm) of 5.5 × $10^6$ cells.

on the rapidity of equilibration between the two tissue pools. The results of the experiments of Loftfield and Harris,[61] and Fern and Garlick,[62] however, suggest that in vivo, equilibrium is reached quite rapidly.

### B. Experimental Procedure

In experiments carried out as outlined in Figure 3 and described in the text, the initial acid-soluble fraction is retained along with the first 5% TCA supernatant fraction obtained from rinsing the precipitate. Unlabeled proline should not be added to the buffer or TCA used in the rinsing procedures. If proline has been used as the radioactive precursor, specific activity in the fraction can be determined quite simply by measuring radioactivity in a portion and using another portion to measure proline by the Troll and Lindsley colorimetric assay procedure.[63] This procedure only can be used, however, if it has been ascertained that proline is the only radioactive component of the fraction. A preliminary experiment should be carried out in which a TCA-soluble fraction is analyzed by cation-exchange chromatography[19–21,43] or any of the chromatographic methods listed for analyzing hydroxyproline.[22–25]

In the case of cation-exchange chromatography, a 1 × 20 cm column of AG 50W-X8 resin (BioRad®) which has been equilibrated with 1 *N* HCl may be used. Elution is achieved with 1 *N* HCl, and 2- to 3-mℓ fractions are collected. The column should be calibrated with a proline solution to determine the position at which the amino acid is eluted, which should be at approximately 90 to 138 mℓ, with the peak at about 104 mℓ. A portion of the acid-soluble fraction is placed on the column, eluted with 160 mℓ of 1 *N* HCl, and a portion of each fraction is counted by liquid scintillation spectrometry. If all of the radioactivity in the acid-soluble fraction is not recovered in the proline peak, then the specific activity should be determined by pooling the fractions in the proline region. HCl is removed by evaporating to dryness under vacuum, the residue is dissolved in a small volume of $H_2O$, and a portion is counted and another analyzed by the procedures of Troll and Lindsley.[63] Specific activity is calculated as dpm per micromole. The amount of collagen synthesis is then calculated as follows:

$$\text{Amount of collagen } (\mu\text{moles}) = \frac{\text{radioactivity in collagen (dpm)}}{\text{proline specific activity } (\text{dpm}/\mu\text{mole})}$$

This value is then divided by the time interval used in order to obtain the absolute rate of synthesis.

## IV. MEASUREMENT OF THE EXTENT OF PROLINE HYDROXYLATION

The 4-*trans*-hydroxylation of up to 50% of the proline residues in the Y position of the Gly-X-Y triplets is essential for the folding of the precursor chains into the correct triple-helical structure of mature collagen.[64] A low hydroxyproline content in the procollagen α-chains brings about instability of the triple helix,[65,66] and as a consequence, collagen secretion may be inhibited.[1,67,68] This in turn may lead to enhanced degradation.[69] In this section the general principles, as well as advantages and disadvantages, of some of the currently used methods for determining the degree of proline hydroxylation will be discussed. These methods fall into either of two major categories: those which quantitate nonradioactive proline and hydroxyproline and those which measure radioactivity in the two imino acids. Finally, we will describe the dual-labeled proline method, a new procedure developed by us to measure proline hydroxylation.[70]

### A. Independent Determination of Nonradioactive Proline and Hydroxyproline

In order to measure the percent proline hydroxylation with these methods, collagen first must be purified and both proline and hydroxyproline content must be determined by specific colorimetric assays or by amino acid analysis.

Stegemann in 1958 introduced a method for hydroxyproline determination based on its oxidation by Chloramine-T,[71] instead of the previously used hydrogen peroxide,[72] yielding pyrrole which reacts with *p*-dimethylaminobenzaldehyde in strong perchloric acid. Several modifications published thereafter[73–76] have increased the sensitivity and specificity of the assay. An enzymatic method to measure hydroxyproline has been reported which relies on the specific oxidation of this cyclic amino acid by a crude enzyme system obtained from *Pseudomonas fluorescens* A-312.[77]

Chinard devised a method based on the formation of a characteristic chromogen when proline interacts with ninhydrin in acidic solutions.[78] Interfering basic amino acids can be removed by adsorption to Permutit, making the method specific for proline.[63]

### B. Simultaneous Determination of Radioactivity in Proline and Hydroxyproline

These methods can be used on acid hydrolyzates of either purified collagen or collagenase released peptides. All of them measure the radioactivities of proline and hydroxyproline when labeled proline is used as a precursor. If specific activities of these amino acids are to be measured, the colorimetric assays described above must be used in conjunction with these methods.

#### *1. Ion-Exchange Chromatography*

Stein and Moore in 1950[20] achieved the quantitative separation of amino acids, including proline and hydroxyproline, by chromatography on columns of sulfonated polystyrene resins (Dowex-50®). This method requires hydrolysis of the sample in 6 *N* HCl as described above. The procedure was proven to be accurate and reliable,[79,80] but nevertheless time consuming and not suitable for rapid processing of a large number of samples. The use of high-speed, high-sensitivity amino acid analyzers has partially overcome that disadvantage.[21] The addition of a stream-splitting device[21] facilitates the measurement of both radioactive and nonradioactive proline and hydroxyproline in the sample.

*2. Chloramine-T Oxidation/Toluene Extraction*

In 1962, Peterkofsky and Prockop introduced a method for the simultaneous measurement of radioactive proline and hydroxyproline in biological materials.[24] It is based on Chloramine-T oxidation of both imino acids and differential extraction of the oxidation products into toluene. $\Delta^1$-Pyrroline, the product of proline, is extracted initially. Upon heating, the hydroxyproline oxidation product, pyrrole carboxylic acid, is converted to pyrrole and then extracted into toluene.[73] Since this procedure is not reproducible unless large amounts of proline and hydroxyproline are added as carriers, modifications of this method were introduced that do not require any carrier.[81,82] In addition, by passing the final toluene extract through a silicic acid column,[83] the contamination of pyrrole by labeled $\Delta^1$-pyrroline can be reduced.

*3. Other Methods of Separation*

Excellent separation of amino acids in acid hydrolyzates, including proline and hydroxyproline, can be achieved with high voltage paper electrophoresis which is based upon the different electrophoretic mobilities of these amino acids.[22,84] A specific color reaction with isatin on the paper can be used for identification and quantitation.[85] This method was reported to be sensitive and rapid and can also be used for the simultaneous determination of radioactive proline, 4-hydroxyproline, and 3-hydroxyproline in biological samples.[86] Identification and quantitation with isatin can also be used after separating proline and hydroxyproline by paper or thin layer chromatography.[23,85]

**C. Dual-Labeled Proline Method**

By simply using a dual-labeled radioactive precursor to label proteins, the collagenase digestion procedure described in Section II.B can be employed as a method for measuring the extent of proline hydroxylation.[70] A decrease in the ratio of $^3H$ to $^{14}C$ in the collagen peptides released will result from specific loss of tritium from the 4-*trans* position of proline during hydroxylation as represented schematically in Figure 5.

*1. General Principles*

The method relies on the following concepts, schematically shown in Figure 6.

1. The ratio of [$^3H$] to [$^{14}C$]proline incorporated into nascent procollagen and nascent noncollagen protein will be identical to that of the isotopes added to the incubation system (4 to 1)
2. During hydroxylation of proline residues (+ ascorbate), tritium at the 4-*trans* position will be displaced from the [4-$^3H$]proline[87]
3. The resultant ratio of $^3H$ to $^{14}C$ in the collagenous peptides (2 : 1) when compared to the initial ratio (4 to 1) will indicate the extent of proline hydroxylation (50%) — If samples are incubated in the presence of $\alpha,\alpha'$-dipyridyl at a concentration which inhibits prolyl hydroxylase activity completely by chelating iron,[88] there

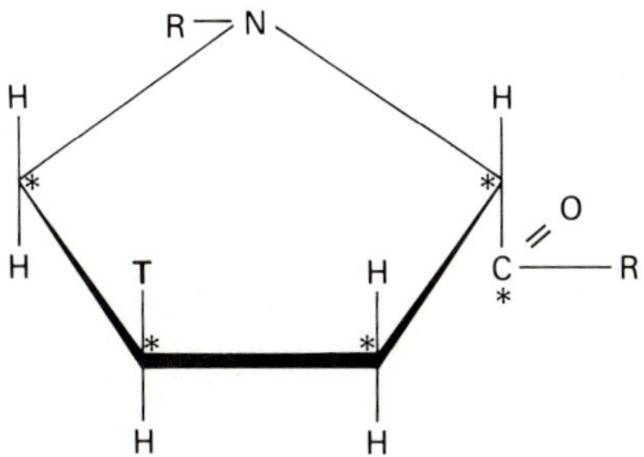


"Pool" of [4-*trans*-$^3$H],[$^{14}$C]-Proline in Nascent Procollagen

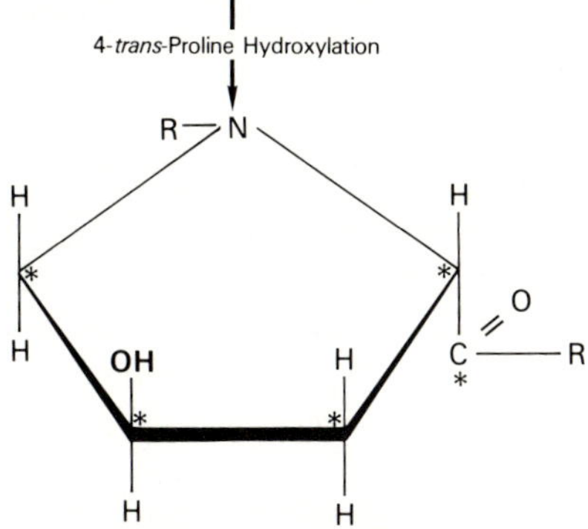


[$^{14}$C]-Hydroxyproline in Collagen

FIGURE 5. Theoretical basis of the dual-labeled proline method to determine proline hydroxylation. The proline molecule shown represents the average of the [4-*trans*-$^3$H], [$^{14}$C]proline pool incorporated into procollagen. T = tritium label; * = $^{14}$C label.

will be no change in the ratio (4 to 1) — Also, the $^3$H to $^{14}$C ratio in noncollagen proteins should remain unchanged, since proline in these proteins does not undergo posttranslational hydroxylation

### 2. *Experimental Procedure*

#### *a. Isotope Mixture*

An isotope mixture that contains [4-$^3$H] and [$^{14}$C]proline is prepared, and unlabled proline is added at concentrations appropriate to the experimental design, as discussed in Section II.B.2.b. In order to minimize the error caused by the spillover of $^{14}$C radioactivity into the $^3$H channel of the liquid scintillation spectrometer, an intial $^3$H to $^{14}$C ratio greater than 4 to 1 should be used. The isotope mixture is lyophilized to remove any tritiated water which may accumulate in the [$^3$H]proline stock solution and is reconstituted in MEM-*O* just prior to use. A mixture containing 10 μCi [4-$^3$H] and 2 μCi [$^{14}$C]proline in MEM-*O* should be adequate for most incubation systems.

#### *b. Incubation Systems*

Tissues or cells and the conditions under which they are incubated are chosen independently of the dual-label method. The method has been satisfactorily used in a variety of systems (e.g., guinea pig calvarial bone, chick embryo frontal bone, and

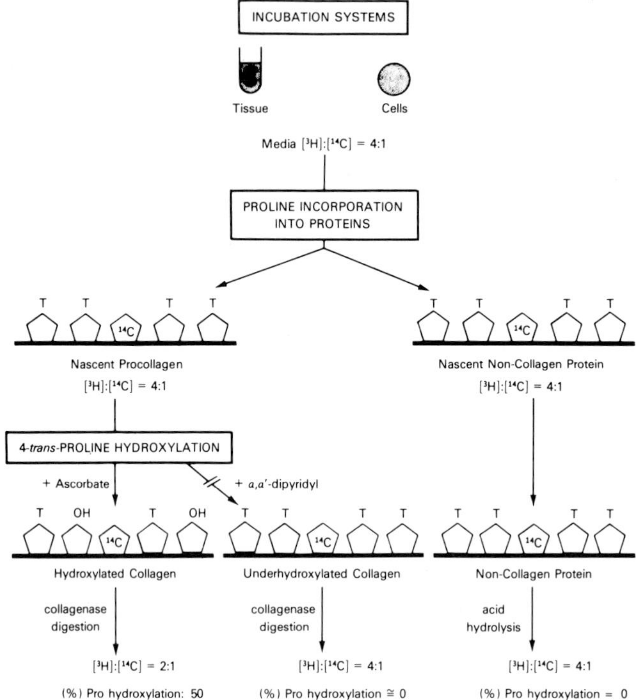


FIGURE 6. Schematic representation of the dual-labeled proline method. Explanatory details in text, Section IV.C.

BALB 3T3, Kirsten sarcoma virus transformed 3T3, and L-929 cells) and under different conditions (addition of ascorbate or α,α′-dipyridyl).

***c. Fraction of Total Tritium in the 4-trans Position of L-[4-$^3$H]Proline***

A tritium atom is displaced from the 4-*trans*, but not from the 4-*cis* position of proline, when either isomer of *L*-[4-$^3$H]proline is converted to 4-hydroxyproline.[87] Although it has been determined that 82 to 84% of the commercially available *L*-[4-$^3$H]proline is labeled in the 4 position, no discrimination between *trans* and *cis* labeling of the 4 position could be made by using nuclear magnetic resonance spectroscopy.[89]

It is indispensable when using the dual-label method to know the fraction of total tritium in the 4-*trans* position of the isotope, and it can be determined in a simple way.[70] Samples are labeled with [4-$^3$H], [$^{14}$C]proline, and collagenase released peptides obtained as described in Section II.B.2. The peptides are hydrolyzed with 6 *N* HCl and proline and hydroxyproline isolated by ion-exchange chromatography, as described above. Tritium remaining in hydroxyproline must be in positions from which it was not displaced during hydroxylation, i.e., at positions other than 4-*trans*. The amount of 3-hydroxyproline in interstitial collagens is negligible, so very little tritium will be displaced from the 3 position. Any tritum not in the 4-*trans* position of hydroxyyproline will also be present in the proline residues of collagen and the amount can be calculated using a simple proportional relationship:

$$[\text{non-4-}trans\text{-}^3\text{H}]\text{proline (dpm)} = \frac{\text{proline}}{\text{hydroxyproline}} \times [^3\text{H}]\text{hydroxyproline (dpm)}$$

The proline to hydroxyproline ratio is determined from the $^{14}$C-radioactivity in the proline and hydroxyproline isolated from the ion-exchange column. The fraction of

**Table 2**
**EXTENT OF PROLINE HYDROXYLATION IN L-929 STATIONARY PHASE CELLS DETERMINED BY DUAL-LABELED PROLINE METHOD[a]**

| Sample number | Ascorbate | Collagenase digestible radioactivity[b] $^{14}$C (dpm) | $^{3}$H (dpm) | $\frac{^{3}\text{H dpm}}{^{14}\text{C dpm}}$ | $^{3}$H to $^{14}$C Ratio in collagenase digest / Initial ratio[c] | Proline hydroxylation[d] (%) |
|---|---|---|---|---|---|---|
| 1 | − | 1957 | 6037 | 3.08 | 0.742 | 37.4 |
| 2 | − | 1729 | 5124 | 2.96 | 0.713 | 41.6 |
| 3 | + | 1713 | 4777 | 2.79 | 0.672 | 47.5 |
| 4 | + | 1904 | 5240 | 2.75 | 0.663 | 48.8 |

[a] Cultures were grown to a density of 2.5 × $10^7$ cells per 100 × 20 mm dish in 15 mℓ of MEM-10. Growth medium was removed and replaced by 3 mℓ of MEM-0 without or with 0.10 m*M* sodium ascorbate. After a 15-min incubation at 37°C under 95% air/5% $CO_2$, 0.2 mℓ of a solution containing 156 μCi [$^{3}$H] and 37.7 μCi [$^{14}$C]proline per mℓ (3.1 m*M*) was added and the incubation continued for another 2.5 hr. The medium was removed, cells were harvested, and protein was digested with collagenase as described in Section II.B.
[b] The difference between the 'plus' and 'minus' collagenase samples.
[c] The initial media ratio of $^{3}$H to $^{14}$C was 4.14.
[d] Calculated by the formula described in the text (Section IV.C).

total tritium present in the 4-*trans* position of *L*-[4-$^{3}$H]proline then is calculated as follows:

$$\text{Fraction in the 4-}trans\text{ position} = 1 - \frac{[\text{Non-4-}trans\text{-}^{3}\text{H}]\text{proline (dpm)}}{\text{Total }[^{3}\text{H}]\text{proline (dpm)}}$$

It was found to be 0.69 for the batch of *L*-[4-$^{3}$H]proline used in the experiments carried out during development of the method and remained constant for at least 6 months.

### *d. Collagenase Digestion*

All procedures used are identical to those described in Section II.B, except that radioactivity of the 5% TCA/0.25% tannic acid-soluble fractions of the (+) and (−) collagenase samples are measured in narrow $^{14}$C and $^{3}$H channels of a liquid scintillation spectrometer. An example of an experiment using this procedure is shown in Table 2. In this experiment, stationary phase L-929 cells were incubated with and without ascorbate. The difference between the $^{14}$C-radioactivities of the (+) and (−) samples is found, converted to dpm, and the $^{14}$C spillover calculated based on the counting efficiency of a $^{14}$C standard in the narrow $^{3}$H channel. After correcting the $^{3}$H cpm in collagen for spillover, the amount of tritium dpm and the $^{3}$H to $^{14}$C ratio in the collagenase digest are calculated.

The initial ratio is determined by diluting the isotope mixture in 5% TCA/0.25% tannic acid and counting a portion equal to the sample volume counted. A background sample, as well as [$^{14}$C]toluene, and [$^{3}$H]toluene standards are prepared correspondingly. The percentage of proline hydroxylation is calculated using the following formula which includes the 4-*trans*-$^{3}$H correction facto (0.69) discussed above.

$$\text{Percent Proline Hydroxylation} = \frac{1 - \dfrac{^{3}\text{H}:{}^{14}\text{C in collagenase digest}}{\text{Initial }^{3}\text{H}:{}^{14}\text{C}}}{\text{Fraction of }^{3}\text{H in the 4-}trans\text{ position}} \times 100$$

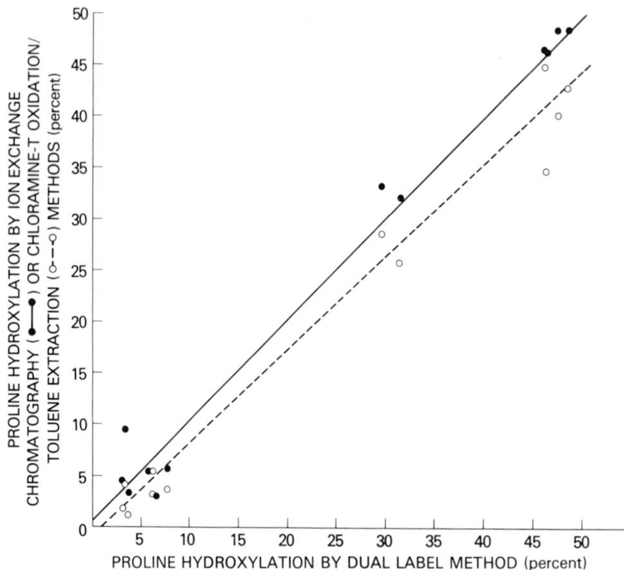


FIGURE 7. Comparison of the dual-label, ion exchange chromatography and Chloramine-T oxidation/toluene extraction methods for determining proline hydroxylation. Each point represents a 2-hr incubation period, at 37°C in 95% air/5% $CO_2$, of either guinea pig calvarial bone or BALB 3T3 cells in logarithmic phase. Bones were incubated in the presence of either 0.5 m*M* sodium ascorbate or α,α′-dipyridyl (0.1 or 0.5 m*M*). Cells were incubated with or without sodium ascorbate (0.1 m*M*) or in the presence of 0.5 m*M* α,α′-dipyridyl. An isotope mixture containing L-[4-$^3$H] and [$^{14}$C]proline was added to the incubation systems as described in Section IV.B. Following collagenase digestion of the TCA precipitates, the percent proline hydroxylation was determined with the dual-labeled proline method. The correlation was very good with both the ion-exchange chromatography ($y = 0.63 + 1.00x$; r:0.99, $p < 0.001$) and the Chloramine-T oxidation/toluene extraction ($y = 0.95 + 0.89x$; $r = 0.98$, $p < 0.001$) procedures.

The same extent of ascorbate-independent proline hydroxylation (82% of maximum) was demonstrated in stationary phase L-929 cells by using the dual-labeled proline method, as had been observed for these cells previously, using the Chloramine-T oxidation/toluene extraction procedure.[10]

The relative rate of collagen synthesis is determined exactly as described in Section II.B.2.e using 6*N* HCl to hydrolyze the noncollagen protein precipitates and using only the radioactivity in the narrow $^{14}$C channel for the calculation.

The $^3$H to $^{14}$C ratio in noncollagen proteins is obtained by counting a fraction of the acid hydrolyzate in the narrow $^{14}$C and narrow $^3$H channels using dual label techniques.

*3. Validity of the Method*

The dual-labeled proline method is simple, reproducible, and rapid, requiring little more time than that necessary for the collagenase digestion. The results in Table 2 show only 1 to 5% variation from the average of the duplicates. The method also has been compared with both the ion-exchange chromatographic and Chloramine-T oxidation/toluene extraction procedures, and a very good correlation with these methods was found (see Figure 7). While determination of the fraction of total tritium in the 4-*trans* position of *L*-[4-$^3$H]proline is required for the method, it need only be carried out once for a given batch of isotope. This procedure may be omitted if the percentage of hydroxylation of one sample relative to another, rather than absolute values, is all that is required. Since the presence of radiochemical impurities in the original isotope solutions will result in an incorrect determination of the initial $^3$H to $^{14}$C ratio, the ratio in either the hydrolyzed noncollagen proteins, or in an $\alpha,\alpha'$-dipyridyl internal assay standard may be used (see Figure 6). In order to measure the percent proline hydroxylation in vivo with this method, the $^3$H to $^{14}$C ratio in the noncollagen protein hydrolyzate of a given tissue sample could be used as the initial ratio.

## V. MEASUREMENT OF COLLAGEN DEGRADATION

### A. Occurrence of Collagen Degradation in Biological Systems

Early studies on collagen metabolism indicated that hydroxyproline excreted in the urine of rats was derived from at least three distinct collagen pools with very different half-lives.[90] It was suggested that at least one of these pools of collagen was degraded very rapidly after its synthesis. Examination of the synthesis and partitioning of hydroxyproline in bone tissue culture[91–93] and in fibroblast cultures[94,95] demonstrated that a large fraction of newly synthesized collagen is degraded. This conclusion was based on finding a low molecular weight fraction containing hydroxyproline in the medium. Recently it has been established in a number of systems that there is a rapid intracellular degradation of a significant portion of newly synthesized procollagen prior to secretion.[79,96] This degradation is not due to extracellular collagenase activity,[79,96] and some agents which increase intracellular cAMP levels cause increased degradation.[97]

The degradation of collagen by collagenase has been recognized as a physiologically important mechanism for regulating extracellular collagen levels,[98] and it is now apparent that intracellular collagen degradation also must play a role in determining the functional levels of collagen. Since the net accumulation of collagen will depend on the difference between the rates of synthesis and degradation, it is important to determine the rate of collagen degradation in order to understand the dynamics of collagen metabolism.

The methods used to quantitate degradation have generally been based on either the detection of the specific products of collagen breakdown, i.e., hydroxyproline containing peptides, or on the measurement of the disappearance of macromolecular collagen species. In this section, we will briefly review the methods currently used to quantitate collagen degradation.

### B. Experimental Procedures

#### *1. Analysis of Hydroxyproline, Free or in Small Peptides*

The catabolism of collagen results in the formation of small hydroxyproline-containing peptides and free hydroxyproline, and a number of studies have relied on the detection of this amino acid in small molecular weight species to determine collagen breakdown. Since hydroxyproline in collagen is formed only by hydroxylation of peptide-bound proline and the hydroxyproline released by catabolism is not reutilized for protein synthesis,[95,99] its appearance in small peptides is an accurate reflection of the degradation of hydroxylated collagen.[100]

Although it is possible to measure collagen degradation by the appearance of hydroxyproline in small peptides chemically,[101] it is usually more informative to

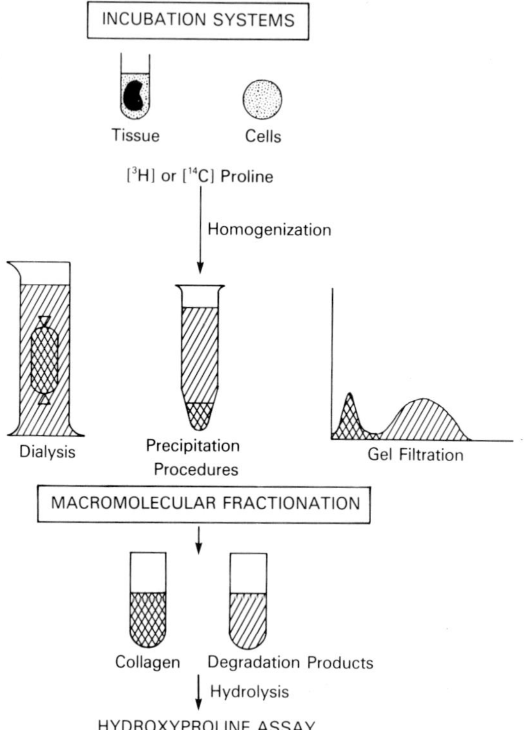


FIGURE 8. Schematic representation of methods used to determine the extent of collagen degradation.

study the degradation of newly synthesized collagen, and this is most readily achieved by using radioactive proline to label the collagen. We will consider methods using radioactive proline to label proteins either continuously or for a short pulse-label followed by examination of the fate of this collagen during a chase period.

After collagen is labeled with either [$^{14}C$] or [$^{3}H$] proline, the undegraded collagen can be separated from the smaller degradation products by a number of methods (see Figure 8), including precipitation of larger molecular weight species with 5% TCA,[102] 67% ethanol,[95] or 0.5 *M* perchloric acid.[94] Gel filtration on Sephadex® G-25 has also been used to successfully fractionate these species,[102] but the most commonly used procedure has been dialysis.[79,93,96,97,102] In this case, the amount of collagen degradation is determined by quantitating the labeled hydroxyproline retained by the dialysis bag and the labeled hydroxyproline that passed through the dialysis bag. Most of the dialysis tubing used have a molecular weight cut off of 10,000 to 12,000, and this corresponds approximately to the lower limit of molecular weights of proteins that are precipitated by cold 5% TCA.[102] These methods can only give an accurate index of collagen breakdown if the molecular size of all the products are less than 10,000 to 12,000 daltons, and thus are able to be separated from the undegraded collagen. Recent studies indicate the validity of these techniques when studying lung explants and fibroblasts in culture.[79,97] Analysis of the products showed that the degradative process produced predominantly small peptides of less than 1000 daltons.

Regardless of the method of fractionation, the high and low molecular weight fractions are acid hydrolyzed and the labeled hydroxyproline content of each fraction is measured by such methods as ion-exchange chromatographic separation of the hydroxyproline or Chloramine-T oxidation followed by extraction of the hydroxyproline oxidation product with toluene. These assay methods are discussed in earlier sections of this chapter (Sections II.A.2, IV.B). After determination of the molecular weight distribution of [$^3$H] or [$^{14}$C] hydroxyproline, the extent of collagen degradation is calculated as the percentage of total hydroxyproline radioactivity appearing in the small molecular weight fraction. It should be recognized, however, that extreme care must be taken when using either of the above hydroxyproline assay techniques. Labeled proline will exist in the small peptide fraction in large excess because of the presence of unincorporated isotope which may be determined artifactually as hydroxyproline. A standard curve of [$^3$H] or [$^{14}$C] proline "spillover" in the hydroxyproline assay must be constructed in order to obtain accurate results.[79,95] Failure to correct for this proline "spillover" will lead to overestimation of collagen degradation, especially when studying systems with low relative rates of collagen synthesis. A further precaution which should be taken is to purify the radioactive proline used for labeling, since some commercial preparations have been found to contain contaminants which elute from the amino acid analyzer in the hydroxyproline region.[103] This can be accomplished by chromatographic separation of the radioactive proline on a Dowex-50® column eluted with 1 *N* HCl and pooling the proline peak (see Section III.B). The use of radioactive lysine to label collagen and subsequent detection of hydroxylysine in the degraded peptides is also a valid method of measuring degradation,[79,93,97] since hydroxylysine, as well as hydroxyproline, is found almost exclusively in collagen.

Using the appearance of low molecular weight species containing hydroxyproline or hydroxylysine as the measure of degradation is advantageous since all degradation subsequent to peptidyl proline or lysine hydroxylation, even rapid intracellular degradation, will be measured. One disadvantage of this approach is that only the degradation of hydroxylated collagen is measured. For the results obtained to be quantitatively valid in systems where hydroxylation is impaired, it must be assumed that the degradation of all the collagenous species present is uniform and this may not be true.[69]

*2. Pulse-Chase Method*

As well as monitoring the appearance of collagen degradation products under continuous labeling conditions, the use of pulse-chase labeling techniques can yield further data on the degradative process. The cell or organ culture is exposed to the labeled proline (or lysine) for a period of time long enough to give adequate incorporation for subsequent analytical procedures, and then this labeling medium is removed and replaced with fresh medium containing unlabeled proline.[79,96] An alternative method to terminate the pulse is to add an excess of unlabeled proline so that the specific activity of the radioactive proline is reduced to such an extent that incorporate of label is reduced to negligible amounts.[52] Incubation of the tissue with chase medium is continued for various lengths of time, again dependent on the characteristics of the particular tissue or cell culture under study. The medium and the cell or tissue compartments are fractionated as described above or macromolecular collagen is measured by collagenase digestion procedure,[35] so that the process of collagen degradation can be progressively monitored. If the pulse time is chosen so that it is less than the time necessary for secretion, then intracellular and extracellular degradation can be determined separately. Secretion begins about 18 min after addition of radioactive amino acid for matrix-free tendon cells,[104] but varies for different types of collagen.[39,105] In lung explants and fibroblasts in culture, this technique indicated that rapid intracellular degradation occurred.[79,96] After a brief labeling period of 20 min, there already had been massive degradation of intracellular collagen, and during chase periods of up to 24 hr when collagen is largely extracellular, there was no further increase in collagen degradation.

## REFERENCES


1. **Blanck, T. J. J. and Peterkofsky, B.,** The stimulation of collagen secretion by ascorbate as a result of increased proline hydroxylation in chick embryo fibroblasts, *Arch. Biochem. Biophys.*, 171, 259, 1975.
2. **Schwarz, R. I. and Bissell, M. J.,** Dependence of the differentiated state on the cellular environment: modulation of collagen synthesis in tendon cells, *Proc. Natl. Acad. Sci. U.S.A.*, 74, 4453, 1977.
3. **Porter, R. R. and Reid, K. B. M.,** The biochemistry of complement, *Nature*, 275, 699, 1978.
4. **Rosenberry, T. L. and Richardson, J. M.,** Structure of 18S and 14S acetylcholinesterase: identification of collagen-like subunits that are linked by disulfide bonds to catalytic subunits, *Biochemistry*, 16, 3550, 1977.
5. **Green, H. and Goldberg, B.,** Collagen and cell protein synthesis by an established mammalian fibroblast line, *Nature*, 204, 347, 1964.
6. **Priest, R. E. and Bublitz, C.,** The influence of ascorbic acid and tetrahydropteridine on the synthesis of hydroxyproline by cultured cells, *Lab. Invest.*, 17, 371, 1967.
7. **Evans, C. and Peterkofsky, B.,** Ascorbate-independent proline hydroxylation resulting from viral transformation of BALB 3T3 cells and unaffected by dibutyryl cAMP treatment, *J. Cell. Physiol.*, 89, 355, 1976.
8. **Nolan, J., Cardinale, G. J., and Udenfriend, S.,** The formation of hydroxyproline in collagen by cells grown in the absence of serum, *Biochem. Biophys. Acta*, 543, 116, 1978.
9. **Bates, C. J., Prynne, C. J., and Levene, C. I.,** The synthesis of underhydroxylated collagen by 3T6 mouse fibroblasts in culture, *Biochim. Biophys. Acta*, 263, 397, 1972.
10. **Peterkofsky, B.,** The effect of ascorbic acid on collagen polypeptide synthesis and proline hydroxylation during the growth of cultured fibroblasts, *Arch. Biochem. Biophys.*, 152, 318, 1972.
11. **Peterkofsky, B., Kalwinsky, D., and Assad, R.,** A substance in L-929 cell extracts which replaces the ascorbate requirement for prolyl hydroxylase in a tritium release assay for reducing cofactor; correlation of its concentration with the extent of ascorbate-independent proline hydroxylation and the level of prolyl hydroxylase activity in these cells, *Arch. Biochem. Biophys.*, 199, 362, 1980.
12. **Mata, J., Assad, R., and Peterkofsky, B.,** An intramembranous reductant which participates in the proline hydroxylation reaction with intracisternal prolyl hydroxylase and unhydroxylated procollagen in isolaed microsomes from L-929 cells, *Arch. Biochem. Biophys.*, 206, 93, 1981.
13. **Feng, J., Melcher, A. H., Brunette, D. M., and Moe, H. K.,** Determination of L-ascorbic acid levels in culture medium: concentrations in commercial media and maintenance of levels under conditions of organ culture, *In Vitro*, 13, 91, 1977.
14. **Mohberg, J. and Johnson, M. J.,** Stability of vitamins in a chemically defined medium for 929-L fibroblasts, *J. Natl. Cancer Inst.*, 31, 603, 1963.
15. **Blanck, T. J. J., Eichhorn, J., and Peterkofsky, B.,** Unpublished results, 1975.
16. **Rowe, D. W., Starman, B. J., Fujimoto, W. Y., and Williams, R. H.,** Differences in growth response to hydrocortisone and ascorbic acid by human diploid fibroblasts, *In Vitro*, 13, 824, 1977.
17. **Hajek, A. S. and Solursh, M.,** The effect of ascorbic acid on growth and synthesis of matrix components by cultured chick embryo chondrocytes, *J. Exp. Zool.*, 200, 377, 1977.
18. **Peterkofsky, B. and Prather, W.,** Cytotoxicity of ascorbate towards cultured cells as a result of hydrogen peroxide formation, *J. Cell. Physiol.*, 90, 61, 1977.
19. **Prockop, D. J., Peterkofsky, B., and Udenfriend, S.,** Studies on the intracellular localization of collagen synthesis in the intact chick embryo, *J. Biol. Chem.*, 237, 1581, 1961.
20. **Stein, W. H. and Moore, S.,** Chromatographic determination of the amino acid composition of proteins, *Cold Spring Harbor Symp. Quant. Biol.*, 14, 179, 1950.
21. **Wiestner, M., Krieg, T., Hörlein, D., Glanville, R. W., Fietzek, P., and Müller, P. K.,** Inhibiting effect of procollagen peptides on collagen biosynthesis in fibroblast cultures, *J. Biol. Chem.*, 254, 7016, 1979.

22. **Formica, J. V. and Katz, E.**, Isolation, purification and characterization of pipecolic acid-containing actinomycins, pip 2, pip 1α, and pip 1β, *J. Biol. Chem.*, 248, 2066, 1973.
23. **Myhill, D. and Jackson, D. S.**, Separation of proline and hydroxyproline using thin layer chromatography, *Anal. Biochem.*, 6, 193, 1963.
24. **Peterkofsky, B., and Prockop, D. J.**, A method for the simultaneous measurement of the radioactivity of proline-$C^{14}$ and hydroxyproline-$C^{14}$ in biological materials, *Anal. Biochem.*, 4, 400, 1962.
25. **Peterkofsky, B. and Diegelmann, R.**, Use of a mixture of proteinase-free collagenases for the specific assay of radioactive collagen in the presence of other proteins, *Biochemistry*, 10, 988, 1971.
26. **Seifter, S. and Harper, E.**, The collagenases, in *The Enzymes*, Vol. 2, 3rd ed., Boyer, P. D., Ed., Academic Press, New York, 1971, 649.
27. **Keil, B.**, Some newly characterized collagenases from procaryotes and lower eucaryotes, *Mol. Cell. Biochem.*, 23, 87, 1979.
28. **Mitchell, W. M.**, The contamination of purified collagenase preparations by clostridiopepidase B (clostripain): the potential effect on studies utilizing collagenase as a highly specific tool, *Johns Hopkins Med. J.*, 127, 192, 1970.
29. **Miyoshi, M. and Rosenbloom, J.**, General proteolytic activity of highly purified preparations of clostridial collagenase, *Connect. Tissue Res.*, 2, 77, 1974.
30. **Lee-Owen, V. and Anderson, J. C.**, The preparation of a bacterial collagenase containing negligible non-specific protease activity, *Prep. Biochem.*, 5, 229, 1975.
31. **Peterkofsky, B.**, Bacterial collagenases, in *Methods in Enzymology, Structural and Contractile Proteins*, Vol. 82, Cunningham, L. W. and Fredriksen, D. W., Eds., Academic Press, New York, 1982, 453.
32. **Nordwig, A. and Strauch, L.**, Reinigung von kollagenase. I. Versuche zur charakterisierung der im rohpraparat vorhandenen begleitenzyme, *Hoppe-Seylers Z. Physiol. Chem.*, 330, 145, 1963.
33. **Bradley, K. H., McConnell, S. D., and Crystal, R. G.**, Lung collagen composition and synthesis, Characterization and changes with age, *J. Biol. Chem.*, 249. 2674, 1974.
34. **Schwarz, R., Colarusso, L, and Doty, P.**, Maintenance of differentiation in primary cultures of avian tendon cells, *Exp. Cell. Res.*, 102, 63, 1976.
35. **Diegelmann, R. F. and Peterkofsky, B.**, Collagen biosynthesis during connective tissue development in chick embryo, *Dev. Biol.*, 28, 443, 1972.
36. **Diegelmann, R. F. and Peterkofsky, B.**, Inhibition of collagen secretion from bone and cultured fibroblasts by microtubular disruptive drugs, *Proc. Natl. Acad. Sci. U.S.A.*, 69, 892, 1972.
37. **Bornstein, P., von der Mark, K., Wyke, A. W., and Ehrlich, H. P.**, Characterization of the pro-α1 chain of procollagen, *J. Biol. Chem.*, 247, 2808, 1972.
38. **Leung, M. K. K., Fessler, L. I., Greenberg, D. B., and Fessler, J. H.**, Separate amino and carboxyl procollagen peptides in chick embryo tendon, *J. Biol. Chem.*, 254, 224, 1979.
39. **Grant, M. E., Kefalides, N. A., and Prockop, D. J.**, The biosynthesis of basement membrane collagen in embryonic chick lens. I. Delay between the synthesis of polypeptide chains and the secretion of collagen by matrix-free cells, *J. Biol. Chem.*, 247, 3539, 1972.
40. **Eichner, R. and Rosenbloom, J.**, Collagen and elastin synthesis in developing chick aorta, *Arch. Biochem. Biophys.*, 198, 414, 1979.
41. **Romanoff, A. L.**, The Avian Embryo, Macmillan, New York, 1960, 990.
42. **Moen, R. C., Rowe, R. W., and Palmiter, R. D.**, Regulation of procollagen synthesis during the development of chick embryo calvaria: correlation with procollagen mRNA content, *J. Biol. Chem.*, 254, 3526, 1979.
43. **Peterkofsky, B. and Prather, W.**, Increased proline transport resulting from growth of normal and Kirsten sarcoma virus transformed BALB 3T3 cells in the presence of $N^6,O^{2'}$-dibutyryl cyclic adenosine 3′,5′-monophosphate, *Arch. Biochem. Biophys.*, 192, 500, 1979.
44. **Oxender, D. L., Lee, M., Moore, P. A., and Cecchini, G.**, Neutral amino acid transport systems of tissue culture cells, *J. Biol. Chem.*, 252, 2675, 1977.
45. **Fasman, G., Ed.**, *Handbook of Biochemistry and Molecular Biology*, 3rd ed., Vol. I., CRC Press, Boca Raton, Fla., 1976, 328.
46. **Antonoglou, O. and Georgatsos, J.**, Studies on a glycoprotein that protects nucleic acids from acid precipitation, *Proc. Soc. Exp. Biol. Med.*, 136, 1360, 1971.
47. **Fessler, J. H. and Fessler, L. I.**, Biosynthesis of collagen, *Annu. Rev. Biochem.*, 47, 129, 1978.
48. **Murphy, W. H., von der Mark, K., McEneany, L. S. G., and Bornstein, P.**, Characterization of procollagen-derived peptides unique to the precursor molecule, *Biochemistry*, 3243, 1975.
49. **Nowack, H., Olsen, B. R., and Timpl, R.**, Characterization of the aminoterminal segment in type III procollagen, *Eur. J. Biochem.*, 70, 205, 1976.
50. **Becker, U., Timpl, R., Helle, O., and Prockop, D. J.**, $NH_2$-terminal extensions on skin collagen from sheep with a genetic defect in conversion of procollagen into collagen, *Biochemistry*, 15, 2853, 1976.

51. **Morris, N. P., Fessler, L. I., and Fessler, J. H.,** Procollagen propeptide release by procollagen peptidases and bacterial collagenase, *J. Biol. Chem.*, 254, 11024, 1979.
52. **Goldberg, B. and Sherr, C. J.,** Secretion and extracellular processing of procollagen by cultured human fibroblasts, *Proc. Natl. Acad. Sci. U.S.A.*, 70, 361, 1973.
53. **Rowe, D., Moen, R. C., Davidson, J. M., Byers, P. H., Bornstein, P., and Palmiter, R. D.,** Correlation of procollagen m-RNA levels in normal and transformed chick embryo fibroblasts with different rates of procollagen synthesis, *Biochemistry*, 17, 1581, 1978.
54. **Breul, S. D., Bradley, K. H., Hance, A. J., Schafer, M. P., Berg, R. A., and Crystal, R. G.,** Control of collagen production by human diploid lung cells, *J. Biol. Chem.*, 255, 5250, 1980.
55. **Olsen, B. R., Guzman, N. A., Condit, C., and Aase, S.,** Purification and characterization of a peptide from the carboxy-terminal region of chick tendon procollagen type I, *Biochemistry*, 16, 3030, 1977.
56. **Hörlein, D., Fietzek, P. P., and Kuhn, K.,** Pro-gln: the procollagen peptidase cleavage site in the α1(I) chain of dermatosparactic calf skin procollagen, *FEBS Lett.*, 89, 279, 1978.
57. **van Venrooij, W. J., Moonen, H., and van Loon-Klaasen, L.,** Source of amino acids used for protein synthesis in HeLa cells, *Eur. J. Biochem.*, 50, 297, 1974.
58. **Airhart, J., Vidrich, A., and Khairallah, E. A.,** Compartmentation of free amino acids for protein synthesis in rat liver, *Biochem. J.*, 140, 539, 1974.
59. **Ilan, J. and Singer, M.,** Sampling of the leucine pool from the growing peptide chain: difference in leucine specific activity of peptidyl-transfer RNA from free and membrane-bound polysomes, *J. Mol. Biol.*, 91, 39, 1975.
60. **Hod, Y. and Hershko, A.,** Relationship of the pool of intracellular valine to protein synthesis and degradation in cultured cells, *J. Biol. Chem.*, 251, 4458, 1976.
61. **Loftfield, R. B. and Harris, A.,** Participation of free amino acids in protein synthesis, *J. Biol. Chem.*, 219, 151, 1956.
62. **Fern, E. B. and Garlick, P. J.,** The specific radioactivity of the tissue free amino acid pool as a basis for measuring the rate of protein synthesis in the rat *in vivo*, *Biochem. J.*, 142, 413, 1974.
63. **Troll, W. and Lindsley, J.,** A photometric method for the determination of proline, *J. Biol. Chem.*, 215, 655, 1955.
64. **Prockop, D. J., Berg, R., Kivirikko, K. I., and Uitto, J.,** Intracellular steps in the biosynthesis of collagen, in *Biochemistry of Collagen*, Ramachandran, G. N. and Reddi, A. H., Eds., Plenum Press, New York, 1976, 163.
65. **Jimenez, S., Harsch, M., and Rosenbloom, J.,** Hydroxyproline stabilizes the triple helix of chick tendon collagen, *Biochem. Biophys. Res. Commun.*, 52, 106, 1973.
66. **Berg, R. A. and Prockop, D. J.,** The thermal transition of a nonhydroxylated form of collagen, *Biochem. Biophys. Res. Commun.*, 52, 115, 1973.
67. **Dehm, P. and Prockop, D. J.,** Synthesis and extrusion of collagen by freshly isolated cells from chick embryo tendon, *Biochem. Biophys. Acta*, 240, 358, 1971.
68. **Peterkofsky, B.,** Regulation of collagen secretion by ascorbic acid in 3T3 and chick embryo fibroblasts, *Biochem. Biophys. Res. Commun.*, 49, 1343, 1972.
69. **Hurych, J., Chvapil, M., Tichý, M., and Beniač, F.,** Evidence for a faster degradation of an atypical hydroxyproline and hydroxylysine deficient collagen formed under the effect of 2,2′-dipyridyl, *Eur. J. Biochem.*, 3, 242, 1967.
70. **Chojkier, M., Peterkofsky, B., and Bateman, J.,** A new method for determining the extent of proline hydroxylation by measuring changes in the ratio of [4-$^{3}$H]:[$^{14}$C]proline in collagenase digests, *Anal. Biochem.*, 108, 385, 1980.
71. **Stegemann, H.,** Mikrobestimmung von hydroxyprolin mit chloramin-T und *p*-dimethylaminobenzaldehyd, *Hoppe-Seylers Z. Physiol. Chem.*, 311, 41, 1958.
72. **Neuman, R. E. and Logan, M. A.,** The determination of hydroxyproline, *J. Biol. Chem.*, 184, 299, 1950.
73. **Prockop, D. J. and Udenfriend, S.,** A specific method for the analysis of hydroxyproline in tissues and urine, *Anal. Biochem.*, 1, 228, 1960.
74. **Woessner, J. F., Jr.,** The determination of hydroxyproline in tissue and protein samples containing small proportions of this amino acid, *Arch. Biochem. Biophys.*, 93, 440, 1961.
75. **Stegemann, H. and Stalder, K. H.,** Determination of hydroxyproline, *Clin. Chim. Acta*, 18, 267, 1967.
76. **Bergman, I. and Loxley, R.,** Two improved and simplified methods for the spectrophotometric determination of hydroxyproline, *Anal. Chem.*, 35, 1961, 1963.
77. **Rosano, C. L.,** Enzymic method for determination of hydroxyproline, *Anal. Biochem.*, 15, 341, 1966.
78. **Chinard, F. P.,** Photometric estimation of proline and ornithine, *J. Biol. Chem.*, 199, 91, 1952.
79. **Bienkowski, R. S., Cowan, M. J., McDonald, J. A., and Crystal, R. G.,** Degradation of newly synthesized collagen, *J. Biol. Chem.*, 253, 4356, 1978.

80. **Stern, B. D., Mechanic, G. L., Glimcher, M. J., and Goldhaber, P.,** The resorption of bone collagen in tissue culture, *Biochem. Biophys. Res. Commun.*, 13, 137, 1963.
81. **Switzer, B. R. and Summer, G. K.,** Improved method for hydroxyproline analysis in tissue hydrolyzates, *Anal. Biochem.*, 39, 487, 1971.
82. **Rojkind, M. and González, E.,** An improved method for determining specific radioactivities of proline-$^{14}C$ and hydroxyproline-$^{14}C$ in collagen and noncollagenous proteins, *Anal. Biochem.*, 57, 1, 1974.
83. **Juva, K. and Prockop, D. J.,** Modified procedure for the assay of $H^3$- $C^{14}$-labeled hydroxyproline, *Anal. Biochem.*, 15, 77, 1966.
84. **Katz, E., and Weissbach, H.,** Incorporation of $C^{14}$-labeled amino acids into actinomycin and protein by *Streptomyces antibioticus, J. Biol. Chem.*, 238, 666, 1963.
85. **Kruze, D. and Wierzchowski, P.,** Determination of traces of proline in biological fluids, *Anal. Biochem.*, 19, 226, 1967.
86. **Tseng, S. C. G., Stern, R., and Nitecki, D. E.,** A new rapid method for quantitating radioactive proline, 4-hydroxyproline, and 3-hydroxyproline, *Anal. Biochem.*, 102, 291, 1980.
87. **Fujita, Y., Gottlieb, A., Peterkofsky, B., Udenfriend, S., and Witkop, B.,** The preparation of *cis* and *trans*-4-$H^3$-*L*-prolines and their use in studying the mechanism of enzymatic hydroxylation in chick embryos, *J. Am. Chem. Soc.*, 86, 4709, 1964.
88. **Cardinale, G. J. and Udenfriend, S.,** Prolyl hydroxylase, in *Advances in Enzymology*, Vol. 41, Meister, A., Ed., John Wiley & Sons, New York, 1974, 245.
89. **Harris, W. G.,** New England Nuclear Corp., Boston, Personal communication, 1980.
90. **Lindstedt, S. and Prockop, D. J.,** Isotopic studies on urinary hydroxyproline as evidence for rapidly catabolized forms of collagen in the young rat, *J. Biol. Chem.*, 236, 1399, 1961.
91. **Golub, L., Glimcher, M. J., and Goldhaber, P.,** The effect of sodium fluoride on the rates of synthesis and degradation of bone collagen in tissue culture, *Proc. Soc. Exp. Biol. Med.*, 129, 973, 1968.
92. **Stern, B., Glimcher, M. J., and Goldhaber, P.,** The effect of various oxygen tensions on the synthesis and degradation of bone collagen in tissue culture, *Proc. Soc. Exp. Biol. Med.*, 121, 869, 1966.
93. **Sakamoto, M., Sakamoto, S., Brickley-Parsons, D., and Glimcher, M. J.,** Collagen synthesis and degradation in embryonic chick-bone explants, *J. Bone J. Surg.*, 61, 1042, 1979.
94. **Steinberg, J.,** The turnover of collagen in fibroblast cultures, *J. Cell Sci.*, 12, 217, 1973.
95. **Steinberg, J.,** Collagen turnover and the growth state in 3T6 fibroblast cultures, *Lab. Invest.*, 39, 491, 1978,
96. **Bienkowski, R. S., Baum, B. J., and Crystal, R. G.,** Fibroblasts degrade newly synthesized collagen within the cell before secretion, *Nature*, 276, 413, 1978.
97. **Baum, B. J., Moss, J., Breul, S. D., Berg, R. A., and Crystal, R. G.,** Effect of cyclic AMP on the intracellular degradation of newly synthesized collagen, *J. Biol. Chem.*, 255, 2843, 1980.
98. **Harris, E. D. and Krane, S. M.,** Collagenases, *N. Engl. J. Med.*, 291, 557, 605, 652, 1974.
99. **Peterkofsky, B. and Udenfriend, S.,** Conversion of proline to collagen hydroxyproline in a cell-free system from chick embryo, *J. Biol. Chem.*, 238, 3966, 1963.
100. **Prockop, D. J. and Kivirikko, K. L.,** Hydroxyproline and the metabolism of collagen, in *Treatise on Collagen*, Vol. 2A, Ramachandran, G. N. and Gould, B. S., Eds., Academic Press, London, 1968, 215.
101. **Boucek, R. J. and Hlavackova, V.,** Synthesis and degradation of collagen in the developing corium of the chick embryo, *Biochem. J.*, 110, 435, 1968.
102. **Holmes, L. B. and Trelstad, R. L.,** Identification of rapidly labeled, small molecular weight hydroxyproline-containing peptides in developing mouse limbs *in vitro* and *in vivo, Dev. Biol.*, 72, 41, 1979.
103. **Berg, R. A., Schwartz, M. L., and Crystal, R. G.,** Regulation of the production of secretory proteins: intracellular degradation of newly synthesized "defective" collagen, *Proc. Nat. Acad. Sci. U.S.A.*, 77, 4746, 1980.
104. **Dehm, P. and Prockop, D. J.,** Time lag in secretion of collagen by matrix-free tendon cells and inhibition of the secretory process by colchicine and vinblastine, *Biochim. Biophys. Acta*, 264, 375, 1972.
105. **Dehm, P. and Prockop, D. J.,** Biosynthesis of cartilage procollagen, *Eur. J. Biochem.*, 35, 159, 1973.

Chapter 3

# IMMUNOCYTOCHEMICAL TECHNIQUES IN CONNECTIVE TISSUE RESEARCH

**F. J. Roll and J. A. Madri**

## TABLE OF CONTENTS

## I. INTRODUCTION

Immunocytochemistry can be defined as the use of labeled antibodies to localize specific antigens in biologic systems. Since Coons et al. introduced antibodies labeled with fluorescent groups in 1941,[1] immunocytochemical methods have provided many insights for biology.

The purpose of this chapter is to describe a number of immunocytochemical techniques that have been successfully applied to connective tissue research. For the most part, they have been previously published in some form, and several are standard techniques used by many laboratories. With two exceptions, we have tried all these methods and are personally familiar with some of their advantages and shortcomings. The methods are accompanied by comments to help readers not familiar with the field decide whether the technique might be useful for their purposes. Clearly such a chapter is selective, and many useful or potentially useful alternative methods have been omitted. The present selection includes ways of looking at connective tissue antigen localization at either the light or electron microscopic level, in whole tissue or in cell cultures.

For a general discussion of immunochemical localization techniques, readers may want to refer to reviews of immunoelectron microscopy by Kraehenbuhl and Jamieson[2] and by Andres et al.,[3] monographs on immunofluorescence microscopy by Goldman[4] and by Nairn,[5] or to a description of immunoperoxidase methods by Sternberger.[6]

There are three absolute requirements for dependable immunocytochemical work: suitable tracers, well-characterized antibodies, and a method of fixation which preserves tissue morphology, but doesn't destroy antigenicity.

As mentioned above, the first tracer used for light microscopic purposes was a fluorescent marker, β-anthracene.[1] Coons and associates soon developed fluorescein conjugates which are the mainstay of light microscopic tracer techniques. Identification of antigens at an ultrastructural level became possible after 1959 when Singer introduced antibody-ferritin conjugates.[7] The third principal type of tracer — horseradish peroxidase — was pioneered by Graham and Karnovksy in 1966.[8] The enzyme was used to label antibody by Nakane and Pierce[9] and by Avrameas and Uriel.[10] This allowed light and electron microscopic visualization of the same sample.

If there has been a lag in applying these immunochemical tracers to connective tissue research, it has been because of the poor immunogenicity of the collagens and the difficulty of making specific antibodies to connective tissue components (see Chapter 11, Volume I). An early paper by Wick et al. was important in this regard for their work showed that unabsorbed rabbit antiserum raised against a highly purified Type I collagen gave a false positive staining in immunofluorescence of renal glomerular basement membranes.[11] Nonimmune sera and affinity-purified antibodies (absorbed on and eluted from a Sepharose®-Type I collagen affinity column) did not give this false positive staining. One can conclude that when the apparently pure immunogen was injected into a rabbit, antibodies were raised to undetected contaminants as well as to the intended immunogen. When the unabsorbed antiserum was used for immunofluorescence, the contaminants or closely related antigens were present in the glomerulus and gave a positive fluorescence. Consequently an immunoabsorption step in which the antiserum is absorbed on and then eluted from purified antigen is generally required in connective tissue immunocytochemical work. In the methods which follow, it is assumed that affinity-purified antibodies are used especially in the first stage of multiple stage procedures (see Chapter 11, Volume I).

When using an immunochemical technique to localize antigens, one tries to fix the sample (tissue or cells) well enough that the antigens of interest and the morphologic appearance are retained, but not so harshly that antigenicity is destroyed. This is a difficult balance to achieve and often requires trial and error in determining a fixation schedule. The results are only reliable if one has good preservation of morphology and appropriate positive and negative controls for each experiment (see below).

The first studies of the connective tissue matrix using purified antibodies were not published until the mid-1970s, but did demonstrate that such techniques could serve as a useful complement to biochemical methods. Questions addressed by these studies included the presence of dermatosparactic collagen (procollagen) in normal tissues,[12] the involvement of the endoplasmic reticulum and Golgi vacuoles of fibroblasts in intracellular processing of collagen,[13] the simultaneous synthesis of collagen Types I and III by the same cell in cultures of human fibroblasts,[14] the expression of different types of collagen during stages of development,[15] and the localization of collagens in normal and pathologic tissues.[16] More recently, immunocytochemical methods have been used to study the association of fibronectin and collagen on cell surfaces,[17] the proposed role of laminin in morphogenesis,[18] and the localization of various connective tissue components in basement membranes.[19-21]

Now that methods are available for generating and characterizing specific antibodies, including hybridoma antibodies, to connective tissue molecules (see Chapters 7, 11, and 12, Volume I), immunocytochemistry will be more widely applied to appropriate problems in connective tissue research.

## II. LIGHT MICROSCOPIC METHODS

### A. Tissue Immunofluorescence

#### *1. Principle*

Conjugates of an antibody and a fluorochrome can be made which upon incubation with frozen tissue sections or with cells in culture dishes will label connective tissue components *in situ*. Structures can be visualized by illuminating the tissue or cells with UV light and observing and recording the fluorescent (longer-wavelength) image emitted.

#### *2. Comment*

An underlying assumption of immunofluorescence and all immunocytochemical techniques is that the antigen of interest in the tissue or cell is accessible to antibody. While experience in this and other laboratories comparing fluorescence localization with biochemical extraction results has shown that this is a valid assumption in general, instances of masking of antigenic determinants are known. For example, when hyaline cartilage, which is known by chemical analysis to contain abundant Type II collagen, is stained with antibody to Type II collagen, little or no labeling can be seen. Presumably this is due to masking of the collagen by the large amounts of proteoglycan in cartilage since von der Mark et al. have shown that pretreatment of cartilage with hyaluronidase reveals Type II staining.[15] Similarly, staining of Type I collagen in bone is facilitated by removing the mineralized matrix with EDTA.[22] Consequently, if we fail to detect a connective tissue component by immunocytochemistry, this can be taken as strong but not conclusive evidence that the component is absent.

As with any method of analysis, there is a limit to sensitivity. Very small amounts of antigen may not be detectable because specific fluorescence does not exceed the background autofluorescence of the tissue or cells. Because biochemical analyses of connective tissue molecules are also limited in sensitivity, false negative immunochemical studies are difficult to prove though they undoubtedly occur. No one has done systematic comparisons of the sensitivities of biochemical and immunochemical techniques for detecting particular connective tissue proteins, but two examples from cell culture studies may give a feeling for the general usefulness of immunochemical methods. In a recent study of amniotic epithelial cells in culture, four connective tissue proteins were detected in the cell layer by biosynthetic labeling.[23] Three of the four were sought by immunofluorescence, and all were detected. The amount of one of these three proteins (procollagen Type III) present in the cell layer was measured by radioimmunoassay and estimated to be less than 1 pg per cell.

In the experience of one of the present authors with biosynthetic labeling and staining of parallel cultures of endothelial or smooth muscle cells, all collagens which were identified biochemically in the cell layer were also detected by immunofluorescent staining.[24] The amount of each type of collagen per culture dish was not determined, but total $^{14}C$-labeled collagenous proteins represent less than 1% of total $^{14}C$-labeled cell layer proteins, so that very small amounts of some types of collagen can be detected. Since immunofluorescence of tissue sections or cell cultures is not a quantitative method, no statements can be made from this type of data about the relative amounts of one or another type of collagen present.

Another caution about immunofluorescence (and other immunocytochemical methods) should be mentioned. That is that while there is a rough correlation between the hemagglutination titer of an antiserum (or antibody) to collagen and its titer by immunofluorescence (i.e., the highest dilution which still gives positive fluorescence), we and others have found that certain antibodies are poorly reactive with tissue despite high titers by hemagglutination or radioimmunoassay. The reason for this phenomenon is unknown, but the consequence is that one may have to screen antibodies from two or three animals to find one that reacts well on tissue sections. Presumably this restriction will apply to monoclonal antibodies as well.

The method below is a version of a standard immunofluorescence technique.[25] The two-stage (sandwich) technique described is more sensitive than a single-stage protocol.[26]

*3. Reagents*

1. Affinity-purified primary antibody to a connective tissue protein prepared as described in Chapter 11, Volume I. The working dilution of a given antibody is determined by staining tissue known to contain the antigen with antibody dilutions from 1 to 10 to 1 to 1000. The highest dilution which still gives strong specific fluorescence is taken as the working dilution. The protein content of nonhybridoma antibodies to collagen at the working dilution in this laboratory is generally 1 to 20 μg/mℓ. Higher concentrations appear to augment nonspecific staining. Individual antibodies may be useful at even greater dilutions, but this must be determined empirically. All antibodies should be kept cold (ice bucket) until they are applied to the tissue. They should be diluted with buffer (PBS) containing 0.1% nonimmune serum or bovine serum albumin (BSA) to stabilize the IgG.
2. FITC (or RITC) conjugated anti-IgG secondary. An affinity purified anti-IgG antibody (e.g., sheep antirabbit) is prepared as described in Chapter 11, Volume 1 and conjugated to a fluorochrome as follows. The method of conjugating fluorescein and rhodamine to IgG used in the laboratory is an adaptation of the methods of Wells et al.[27] and The and Feltkamp.[28] A 1 mg/mℓ solution of 5-fluorescein isothiocyanate (FITC) (Eastman Kodak Company®) or rhodamine B-isothiocyanate (RITC) (Sigma®) in 0.15 *M* $Na_2HPO_4$-2 $H_2O$, pH 9.0, is made up immediately prior to use. Affinity-purified IgG in PBS (5 mg/mℓ) is pH-adjusted to 9.5 with a 0.1 *M* $Na_3PO_4$-10 $H_2O$ and mixed with the FITC or RITC solution at a 1 to 100 mg dye to milligram IgG ratio at room temperature (22°C) for 1 hr. The reaction is then quenched by the addition of an equal volume of 1% ethanolamine (Sigma) in 0.15 *M* $Na_2HPO_4$-2 $H_2O$, pH 9.0. After stirring the mixture for 15 min at room temperature, the solution is passed over a BioGel® P-2 column equilibrated in PBS to separate unreacted dye from conjugate (see Figure 1). The conjugate can be stored frozen at −70°C in 1-mℓ aliquots. Repeated freeze-thawing should be avoided as this will cause protein denaturation. The F/P ratio can be determined from the $OD_{280}$ and $OD_{493}$ values using a nomogram (see Figure 2). In our hands, the optimal F/P ratio is between 3 and 4. The titer of the conjugate can be tested using standard Ouchterlony plate

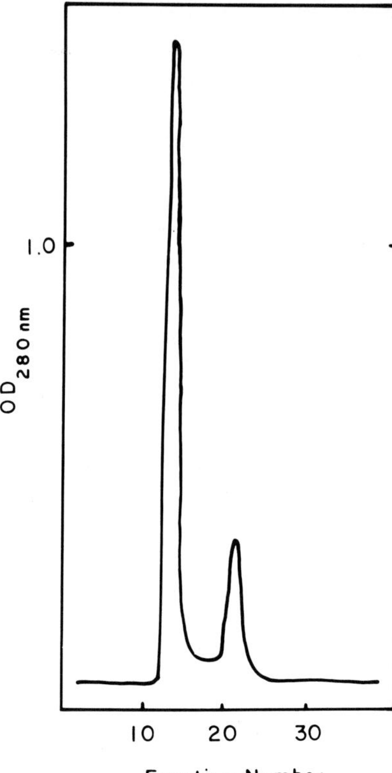


FIGURE 1. (P-2 column) Elution profile of FITC-IgG conjugate from P-2 column as described in Section II.A. The conjugate elutes in the void volume (tube number 13 to 17), while the unreacted dye elutes in the included volume of the column (tube number 20 to 24).

double diffusion techniques, however, each conjugate should be titered by tissue immunofluorescence to determine the working dilution. The working dilution is determined empirically as described for the primary antibody. FITC-conjugated secondary antibodies are generally used at a protein concentration of 100 to 500 μg/mℓ. They can be stored concentrated in plastic centrifuge tubes, and aliquots can be thawed on the day of use. Once diluted they should probably be used within a few days. PBS is an adequate diluent. BSA (0.1%) is added to the PBS diluent to stabilize the IgG if it is to be stored at 4°C for more than a day. For obvious reasons, nonimmune serum of the species against which the secondary is directed cannot be used in the diluent. Individual batches of FITC-conjugated antibody may form microscopic precipitates on storage. These can adhere to the section and cause artifacts ("snow"). If this is a problem, the secondary antibody can be centrifuged (39,000 × g) for 20 min just before use to remove the precipitates.

3. Phosphate buffered saline (PBS), pH 7.4.
4. Tissue-Tek II O.C.T. Embedding compound, Lab-Tek Products, Inc., Division Miles Laboratories®, Napierville, Ill.
5. Polyvinylalcohol-glycerol mounting medium. This is prepared by dispersing 20 g of Gelvatol® (Monsanto®, Bircham Bend Plant, Indian Orchard, MA) in 80

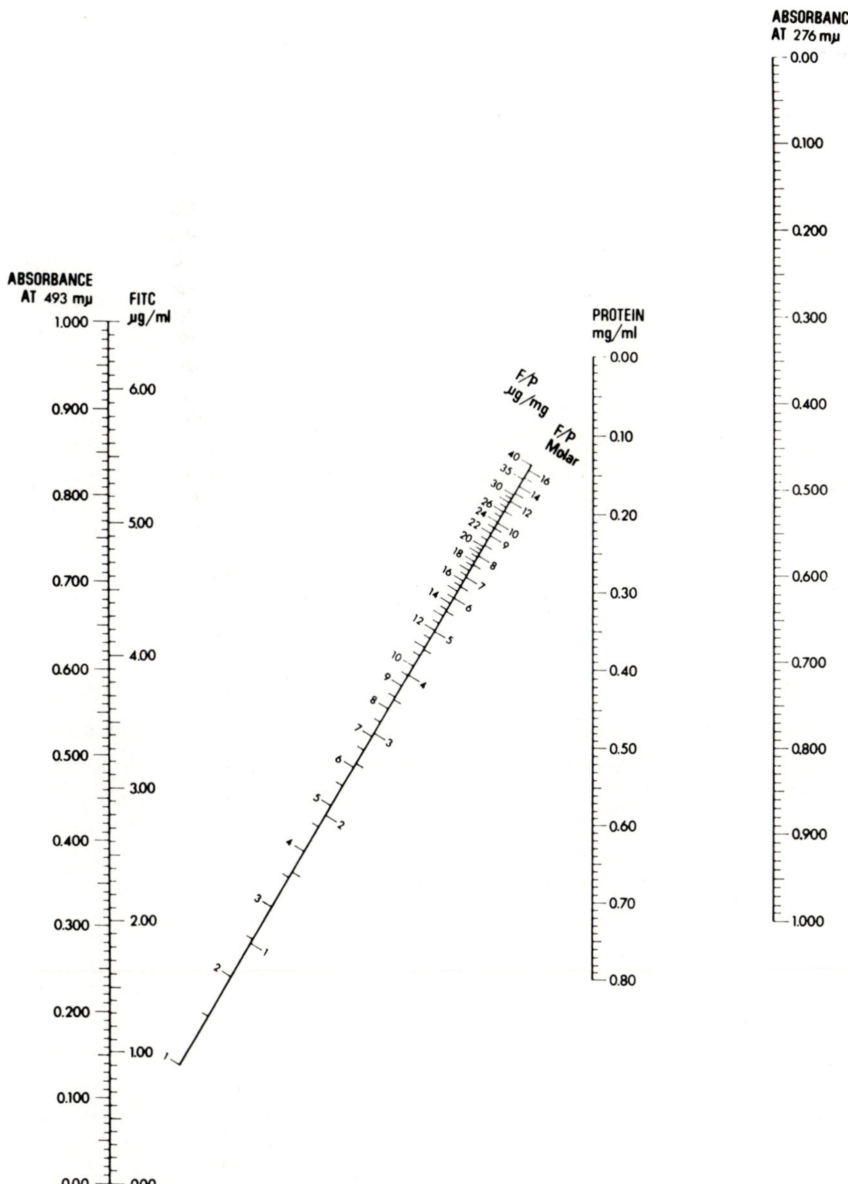


FIGURE 2. Nomograph for the determination of FITC and protein in FITC-globulin conjugates at pH 7.2. A dilution of the conjugate with PBS is chosen which has an absorbance of 0.3 to 0.5 at 275 mμ (about 1 to 100 dilution), The absorbance at 275 mμ is plotted on the right ordinate, and the peak absorbance at 493 mμ is plotted on the left ordinate. Note that the peak absorbance may vary slightly from 493 mμ. To read the F/P ratio, a line is drawn between these two points. The FITC and protein concentration of the diluted sample can be similarly read. (From Wells, A. F., Miller, C. E., and Nadel, M. V., *Appl. Microbiol.*, 14, 271, 1966. With permission.)

mℓ of PBS, pH 7.2, on a hot plate to dissolve the resin. After the solution cools, 40 mℓ of glycerol is added. This solution (pH 6 to 7) is stored at 4°C.

6. Acid alcohol slide wash prepared by mixing 10 parts concentrated HCl with 990 parts 70% ethanol.
7. Albumin for coating slides. Albumin can be inexpensively made from the white

of an egg added to one tenth volume of glycerol. This solution can be stored at 4°C for months.

8. Acetone, absolute, for fixation.
9. Inhibitor solutions. Solutions of connective tissue proteins to be used as inhibitors in immunofluorescence controls are made as follows: the protein is dissolved in a neutral buffer such as 50 m*M* Tris-150 m*M* NaCl — 0.1% nonimmune serum at a concentration of 200 μg/mℓ on the day of the experiment. If the antigen is not readily soluble in a neutral buffer, it can be dissolved in 0.05% acetic acid at 1 mg/mℓ and diluted with the neutral buffer to 200 μm/mℓ. The pH can be adjusted with small aliquots of 1 *M* Tris base if necessary.

### 4. *Preparation of Tissue*

Preservation of morphology is better in fresh tissue than in tissue taken at autopsy or hours after death. In addition, there is a possibility of post-mortem degradation of connective tissue antigens and of nonspecific absorption of IgG to necrotic cells — all reasons to work with the freshest available tissue. Sometimes this may require using an animal model.

Fresh tissue is cut with a razor blade into pieces of a convenient size for cryostat sectioning: a block 1 $cm^2$ and 2 to 3 mm thick. A few drops of O.C.T. compound are poured onto a conventional metal cryostat chuck, and the tissue block is laid flat on the surface of the O.C.T. More O.C.T. is poured on the upper surface of the tissue, and the chuck and tissue are partly immersed in liquid $N_2$ (−196°C). Having a generous margin of O.C.T. compound around the specimen will make sectioning easier. If the liquid $N_2$ is kept in a wide-mouth thermos jar, the chuck and specimen can be lowered in upright by means of a wire holder. During quick freezing, the upper surface of the tissue should be just at or slightly below the surface of the liquid $N_2$. If the chuck is immersed too deeply at first, the O.C.T. will crack or bubble, making sectioning more difficult. When the chuck has reached the temperature of the liquid $N_2$, it can be removed to the cryostat chamber and allowed to come to cutting temperature (usually −25 to −30°C, but may vary with the tissue) over a period of an hour. If sectioning is to be delayed, the embedded tissue can be stored in an airtight plastic bag at −30°C for several days. For prolonged storage (up to a year), tissue blocks without O.C.T. can be wrapped in aluminum foil and quick frozen and stored in liquid nitrogen. If this is not available, quick freezing followed by storage at −70°C in an air-tight container is an alternative. We have found that tissues stored for a year at −30°C in plastic bags sealed with tape lose normal morphologic appearance and have altered immunofluorescent staining patterns with an increase in nonspecific staining.

### 5. *Preparation of Slides*

Frosted glass microscope slides are cleaned by dipping them in acid alcohol (see Reagents section above) and drying them. The slides are then coated with a very thin film of albumin (see Reagents section above) to increase adhesion of tissue sections. This is especially important with delicate tissues such as lung and skin. The albumin is spread thinly over one side of the slide with a fingertip. The albuminated slides are baked in a drying oven at 100°C for 30 min and then brought to room temperature for use.

### 6. *Sectioning*

Frozen sections (6 to 8 μ thick) are cut in the cryostat. The cutting temperature, knife blade angle and sharpness, and the ambient humidity all influence the quality of the sections. At too low a temperature, the tissue will fracture and shred, while at too high a temperature, the sections will curl up and stick to the knife blade. For most tissues, a temperature of −25 to −30°C is satisfactory. The optimal knife angle depends on the type of instrument and can be empirically determined. Low humidity is desirable to prevent curling of the sections on the knife blade and distortion of tissue morphology.

As sections come off the block, they can be helped onto the knife blade with a small artist's brush. The sections are picked up from the blade by touching the albumin-coated surface of a glass slide to the section. The slides are immediately placed in an airtight box in a refrigerator. Sections will adhere better if they are stored in this way overnight and fixed and stained the following day. If staining is delayed for several days to a week, the sections should be fixed (see below) the day after they are cut and stored at 4°C.

*7. Immunofluorescent Staining*

Slides are brought to room temperature while still in an airtight container (especially if humidity is high). All subsequent steps of the staining procedure can be done at room temperature. A circle is scored with a diamond scribe around the sections to hold reagent drops in place. Then the tissue is fixed by immersing the slides in a Coplin jar of acetone for 10 min. Slides are air-dried for 5 min and then rehydrated in PBS for 5 min. Throughout the rest of the procedure, the sections should never be allowed to dry. For this reason, it is convenient to handle no more than a dozen slides at a time and to keep them in a moisture chamber during incubations.

The sections are first overlaid with 50 to 75 $\mu\ell$ of the primary antibody diluted appropriately (see Reagents section above) with PBS containing normal serum and 0.02% sodium azide. Sections are incubated with primary antibody for 1 hr. They are then washed three times with PBS for a total of 10 min, either in a Coplin jar or by gently flooding them with a Pasteur pipette. After the last wash, the slide is wiped dry except for the area in the enscribed circle containing the section. A drop (50 to 75 $\mu\ell$ of appropriately diluted FITC-conjugated secondary antibody (see Reagents section above) is applied. Slides are incubated for a further 50 min. Exposure to direct light which might cause fading of the fluorescence can be prevented by covering with a foil-lined top.

The sections are again washed three times with PBS for a total of 10 min. The slides are dried around the scored area and a drop of PVA (see Reagents section above) is placed over each section. A coverslip (#1 thickness) is mounted over the section with care to avoid trapping air bubbles over the section. Slides can be viewed immediately with a fluorescence microscope, but it is better to let the PVA medium set first by keeping the slides at 4°C for an hour. The stained sections are stable stored in the dark at 4°C for 1 to 2 weeks, or longer, before fading of fluorescence occurs.

*8. Controls*

Staining of tissues can be either specific, as when antibody reacts with the intended antigen in the tissue and the antibody is in turn recognized by a labeled secondary antibody, or it can be nonspecific, as when one of the reagents adheres to some other component than the intended antigen. To avoid such pitfalls every cytochemical study includes specificity controls:

1. *Autofluorescence control*—PBS is substituted for both primary and secondary antibodies.

2. *FITC-conjugate control*—PBS is substituted for the primary antibody. Positive fluorescence indicates nonspecific adhesion of the FITC conjugate. See References 4 and 5 for a discussion of the problem.
3. *Nonimmune serum control*—A comparable concentration of a homologous nonimmune serum or IgG is substituted for the primary antibody. If positive fluorescence occurs, labeling with the specific antibody is suspect. See References 4 and 5 for a discussion.
4. *Antigen inhibition control*—The primary antibody is preincubated with small amounts of purified antigen (the inhibitor). If the antibody is specific, its antigen combining sites will be occupied by the inhibitor and the antibody should no longer stain tissue. In practice purified primary antibody at twice the working concentration is incubated with an equal volume of inhibitor solution (see Reagents section above for 30 min at room temperature. The solution is centrifuged at 39,000 × g for 20 min to remove any large precipitation which might form. The supernatant (upper ⅔ of solution) is pipetted off carefully and used in place of the primary antibody. In such an inhibition experiment, the specific antigen, but not other antigens, should inhibit fluorescent staining. Some residual "snow" due to small antigen-antibody complexes is to be expected in this control.

*9. Viewing Slides*

To discuss the various types of instruments available for fluorescence microscopy or to list the pitfalls in interpretation are beyond the scope of this chapter. Both topics are discussed in excellent books by Goldman[4] and by Nairn.[5] The essentials for immunofluorescence are good-quality objectives, a lamp of the mercury, xenon, or halogen type, and specific excitation and barrier filters for fluorescein and rhodamine, as well as a camera attachment.

We have had experience with a Zeiss Standard WL Research Microscope® equipped for immunofluorescence with a vertical illuminator (epi-illumination), 10, 25, and ×40 Neofluor® objectives (N.A. 0.30, 0.60, and 0.75, respectively), as well as a ×40 Planapochromat objective (N.A. 0.95). The light source is a 50-W high-pressure mercury lamp, and the filters include a 450- to 490-nm excitation and 520-nm barrier filters for fluorescein and 515- to 560-nm excitation and 590-nm barrier filters for rhodamine. These allow double immunofluorescence labeling experiments on the same section.

One can view stained slides as soon as the PVA medium has set. The light source is turned on a few minutes before viewing, but light intensity will not be stable enough for photography until the lamp has had a longer time to warm up (see manufacturer's information). The lamp is usually turned on only once a day no matter how long it is to be used, since repeated switching on and off shortens bulb life considerably. The average bulb of the above type lasts 100 to 200 hr.

When viewing stained tissues, one quickly gains an appreciation for the rapid fading of the fluorochrome. This is due to irreversible formation of nonfluorescent thermal breakdown products of the fluorochrome as well as to reversible changes in its energy level. If one views an area for 30 sec with a high-magnification (×40) objective and epi-illumination, the area is visibly bleached. Consequently one has to learn to focus and choose an area for photography quickly. Some of the original fluorescence intensity returns if the sections are kept in the dark at 4°C for days to weeks.

*10. Interpretation*

Since experience is necessary in distinguishing true fluorescein immunofluorescence (apple-green) from autofluorescence (blue-green or yellow) and in recognizing such artifacts as "snow" (microaggregates of FITC-conjugated antibody) or physically trapped conjugate at the edge of a section or beneath a fold in the tissue, interpretation is easier if one begins by studying staining of a tissue in which the distribution of the antigen is reasonably well known. In this way, one can not only learn to recognize specific fluorescence, but also work out the working concentration of primary and secondary antibodies to be used in studying new tissues. For example, kidney tissue contains several interstitial and basement membrane collagens with unique patterns of distribution in our experience,[19] so we use frozen sections of kidney from the mouse to determine the best working concentration for each new batch of antibody or FITC-conjugate.

*11. Photography*

The intensity of light emitted by an FITC-labeled frozen section is very low in comparison with the light available for conventional photomicrography. As a result, exposure times are longer and film with a high ASA rating is preferable. With the equipment described above and Kodak Ektachrome Color Slide Film® (ASA 400), we routinely use 4-min exposure times. With different objectives and light sources, exposure times will differ and can probably be shortened. To increase apparent intensity of positive fluorescence, this film can be pushed to 800 during processing without much loss of resolution (see Figure 3). An exposure meter attached to the camera is not necessary, since using a standard exposure time allows one to compare the relative intensities of staining of different antibodies.

Because of the expense and difficulty of reproducing color photographs for publication, black and white negatives are often preferable. They can be made using Kodak Recording Film® with comparable exposure times. This film can be developed so as to give an effective ASA rating of 4000 without excessive graininess. Alternatively, black and white prints can be made from the Ektachrome Color Slides®.

**B. Cell Culture Immunofluorescence**

*1. Principle*

Refer to previous procedure in Light Microscopic Methods.

*2. Comment*

Nearly the same procedure as that outlined above for frozen sections can be followed for cultured cells, the only differences stemming from the fact that the cells have been grown on plastic or coverslip glass substrata so that the preparations for staining are more simple.

Cells maintained in almost any size of plastic culture dish or flat-bottomed flask can be stained for immunofluorescence by cutting the plastic with cells attached to a convenient size and, after staining, viewing with epi-illumination. We have used dishes made by Costar®, Falcon® and Corning®, but one of the authors, Joseph Madri, has found that Thermanox® plastic coverslips made by Lux Scientific Corporation, and used by some as a cell culture substratum, cannot be used for immunofluorescence work because they are brightly autofluorescent. This is not a problem with glass coverslips.

A potential problem with cells grown on collagen gels[29] or extracellular matrix preparations[30] is cross-reactivity of the exogenous matrix, with the primary antibody leading to false positive staining. In part one can avoid this by using matrix material which has been tested for cross reactivity (e.g., by immunofluorescence). However, one cannot exclude proteolytic degradation of the matrix during culture. Antigenic

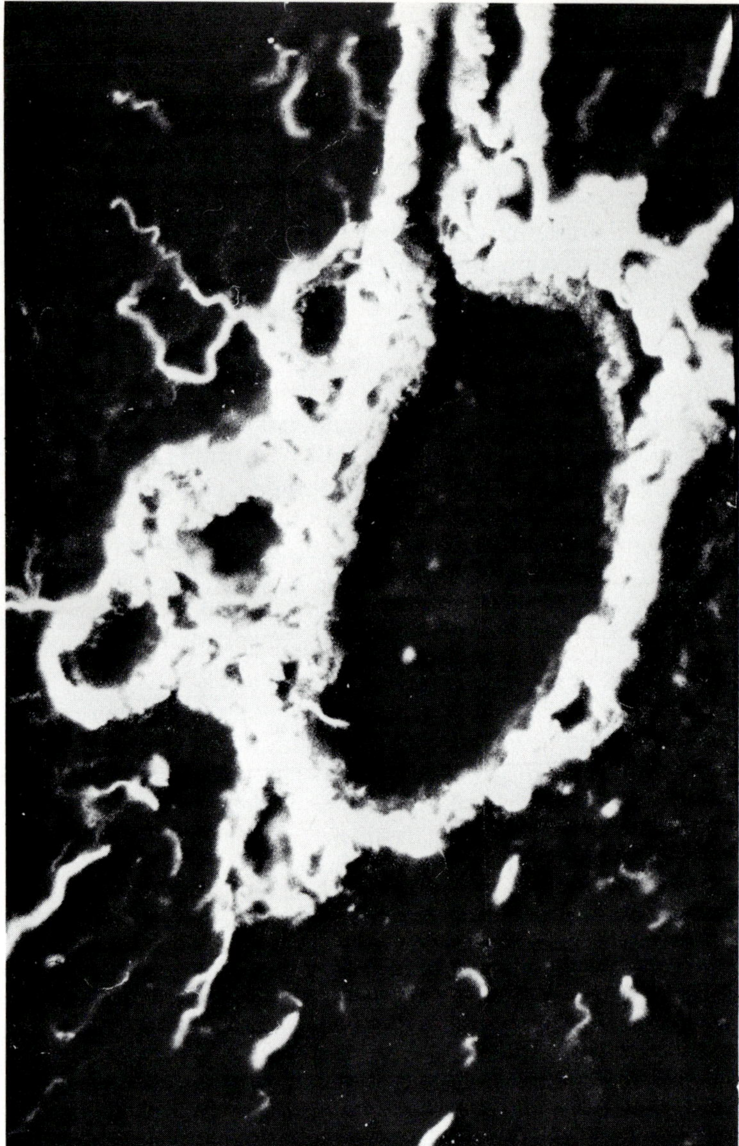

FIGURE 3. Tissue immunofluorescence. Frozen section of rat liver showing the dense connective tissue bundles of a portal triad stained for Type III collagen. As described in the text (see Section II.A), the tissue was fixed with acetone for 10 min and incubated with affinity-purified rabbit antibody to rat Type III collagen (10 μg/mℓ and then with FITC-conjugated sheep antirabbit IgG (100 μg/mℓ; magnification ×250.

sites in the matrix might be exposed and cause false positive staining. Therefore, ideally, the presence or absence of the antigen of interest in the matrix should be known in order to interpret such a study.

The longer the interval between plating and staining a dish of cells, the more time the cells have to produce an insoluble matrix and the more pronounced the immunofluorescence staining of matrix components will be. In general it is best to let the cells at least achieve confluency before staining, but this will depend upon the nature of the experiment. Other general considerations are whether to use ascorbate to increase collagen synthesis[31] and whether to use β-aminoproprionitrile to decrease cross-linking. The effect of the latter might be to keep newly synthesized collagen in the medium by inhibiting lysyl oxidase induced cross-linking. Increased solubility of collagen incubated with β-aminopropionitrile and lysyl oxidase as opposed to collagen incubated with lysyl oxidase alone was shown by Siegel and Fu.[32]

### *3. Procedure*

#### ***a. Preparation of Cells***

The medium is decanted and the cell layer is washed three times with PBS for a total of 10 min.

#### ***b. Fixation***

After the third wash is decanted, the cells are fixed for 10 min with absolute methanol at room temperature. Acetone will dissolve some plasticware, and we do not use it for this reason. Solvents such as acetone and methanol will open up cell membranes and allow immunochemical reagents access to the interior of the cell so that intracellular as well as extracellular labeling occurs. If labeling of just cell surface and matrix is desired, cells should be fixed with 2% paraformaldehyde in PBS, pH 7.4 for 15 min, followed by three washes in 0.1 *M* glycine −0.05 *M* Tris®, pH 7.4 for a total of 10 min.[17] We have not had experience with immunofluorescence labeling of cells sealed in this way, but the method is commonly used.

### *4. Labeling Procedure*

There are a number of ways of handling culture dishes for staining, depending upon the size of the dishes and how many experiments must be done with each. Small (35 or 50 mm) plastic culture dishes can be stained without subdividing them. Alternatively, these or larger dishes can be cut into convenient sizes with a red hot scalpel or wire. Cultures on glass coverslips can be supported on a hard piece of plastic and manipulated with fine forceps.

If one is using cells grown on plastic surfaces, it is helpful to score a circle in the plastic (1 cm in diameter) to keep reagent drops in place. All the staining steps are done exactly as described for frozen sections (see Reagents section above) Washing is conveniently done by flooding the dish or coverslip with PBS and gently aspirating it with a Pasteur pipette connected to a vacuum aspirator. After immunostaining, the piece of plastic or glass bearing the cells is coverslipped, viewed, and photographed in the same way as are frozen sections.

### *5. Controls*

The same controls are used for cell culture immunofluorescence as are described for tissue immunofluorescence (see previous section).

## C. Tissue Immunoperoxidase

### *1. Principle*

Enzymes with peroxidatic activity (e.g., horseradish peroxidase, a glycoprotein with a molecular weight of 40,000 daltons) can be coupled to IgG without marked loss of enzyme activity.[33] The enzyme-IgG conjugate can be localized in immunocytochemical studies by the addition of $H_2O_2$ and the substrate diaminobenzidine (DAB). The substrate will be oxidized to a brown, electron-dense, insoluble reaction

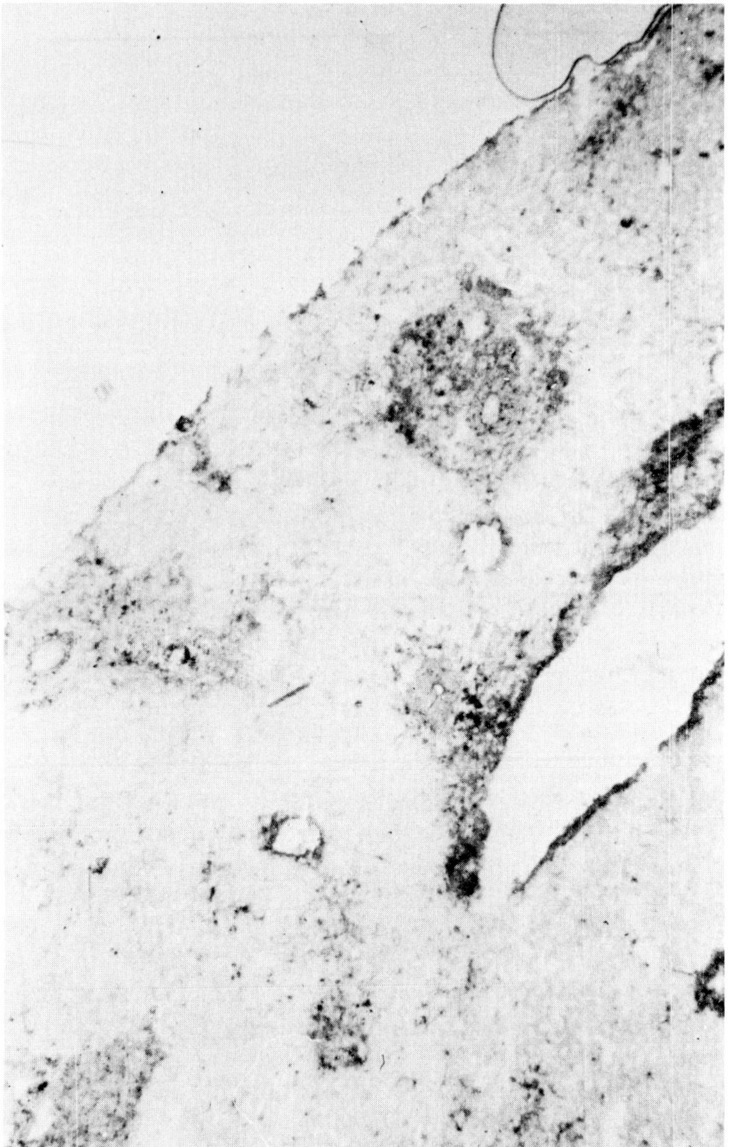

FIGURE 4. Tissue immunoperoxidase. Frozen section of mouse liver with Schistosome granuloma stained for Type III collagen by a sandwich technique. The primary antibody is rabbit antimouse Type III collagen, and the secondary is a sheep antirabbit IgG conjugated to horseradish peroxidase as described in Section II.C; magnification ×100.

product which will be deposited primarily in the vicinity of the enzyme-IgG. The reaction product can be seen at both a light and electron microscopic level (see Figures 4 and 5).

*2. Comment*

Labeling with peroxidase-conjugated reagents has several advantages over labeling with fluorescent tracers. One such advantage is the amplification inherent in the enzyme-catalyzed conversion of substrate to electron-dense, chromogenic product.

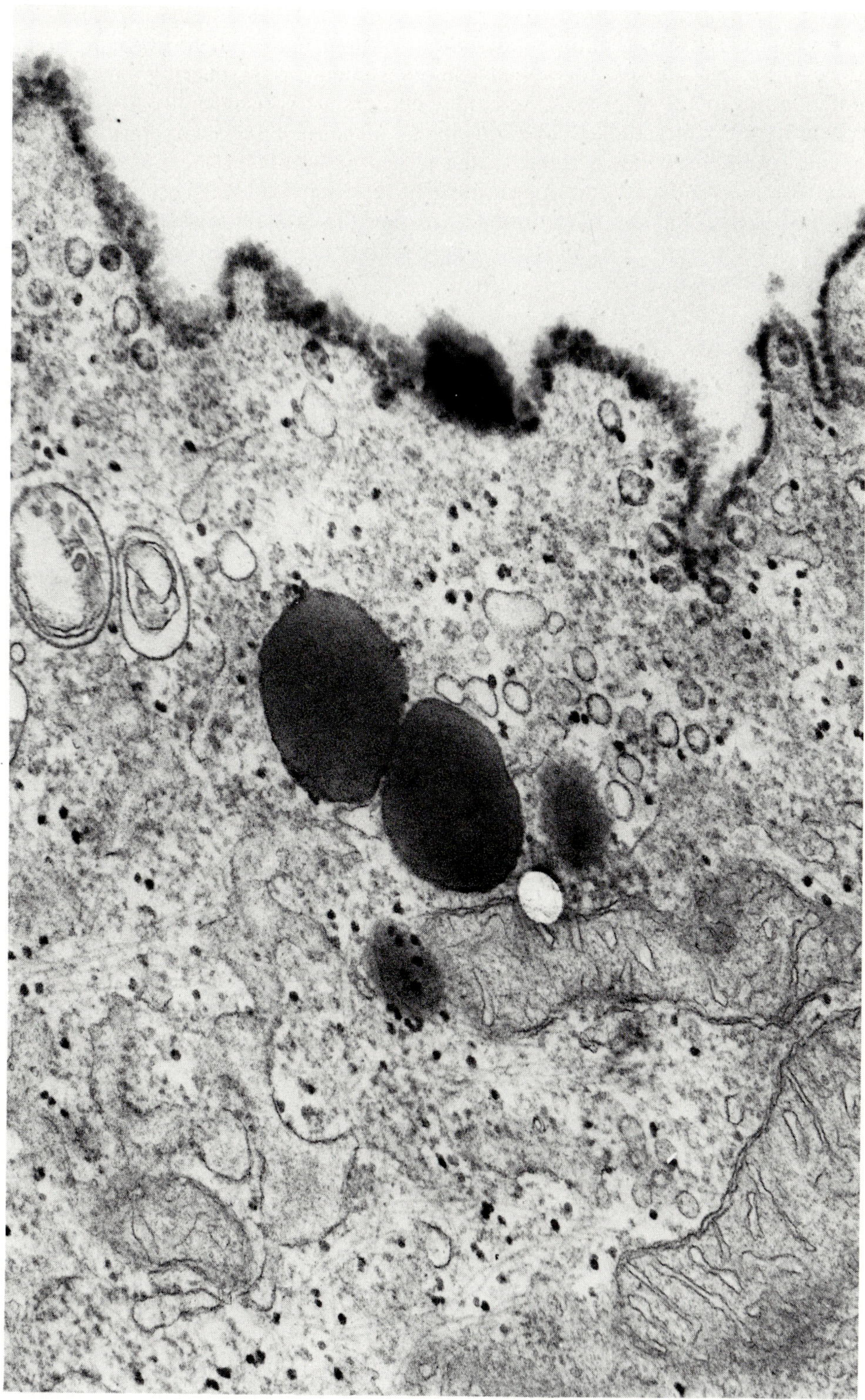

FIGURE 5. Electron micrograph of an endothelial cell culture labeled with antibody to Type V (AB) collagen and a horseradish peroxidase conjugated secondary antibody. Note the uniform black reaction product on the cell surface; magnification ×156,000.

Another advantage over fluorescent tracers is that the labeled tissue section or cell culture can be preserved almost indefinitely without fading of the chromogen and can be visualized and photographed using a standard light microscope. Another significant advantage of peroxidase-conjugated probes is their ability to be used in electron microscopic studies. The drawbacks of peroxidase labeling are primarily that fine resolution is not possible at the ultrastructural level because of diffusion of the reaction product[34] and that in most tissues there is less contrast between a labeled structure and the background than there is in a fluorescence study. The problem of interference by endogenous peroxidases must also be kept in mind when one is using this technique.

*3. Reagents*

1. Horseradish peroxidase-IgG conjugate (adopted from method of Nakane and Kawaoi[33]) — Horseradish peroxidase (Sigma HRPO, type VI, E.C. 1.11.1.7) is dissolved in fresh 0.3 *M* sodium bicarbonate, pH 8.1, at a concentration of 5 mg/mℓ. Fluorodinitrobenzene (1%) (Sigma) is added with gentle stirring at a 1:10 HRPO to FDNB (v/v) ratio. After 1 hr at room temperature (22°C), 0.08 *M* $NaIO_4$ in double distilled water is added at a 1 to 1 HRPO to $NaIO_4$ (v/v) ratio. The solution is stirred gently for 30 min at room temperature, during which time the color of the brownish solution will become green-yellow. After this, 0.16 *M* ethylene glycol in double distilled water is added at 1 to 1 HRPO to ethylene glycol (v/v) ratio, and the solution is stirred gently for 1 hr at room temperature. The solution is then dialyzed against four 2-ℓ changes (12 hr each) of 0.01 *M* sodium bicarbonate buffer, pH 9.5, at 4°C. The dialyzed activated HRPO solution is then added to affinity-purified IgG dissolved at a 1 to 1 HRPO to IgG (w/w) ratio. After gently stirring the solution for 3 hr at room temperature, sodium borohydride is added in a 1 to 1 HRPO to $NaBH_4$ (w/w) ratio. The solution is then stirred gently overnight at 4°C. The stabilized conjugate is then dialyzed against PBS (do not use azide as an antibacterial agent as it will inhibit the enzyme), concentrated using Aquacide (Cal-Biochem®) or an Amicon ultrafiltration system, and centrifuged at 5000 r/min for 30 min at 4°C to remove any precipitate. The conjugate is then chromatographed on a P-150 column (2.5 × 90 cm) equilibrated in PBS to separate the HRPO-IgG conjugate from free HRPO (see Figure 6). The conjugate can be titered by Ouchterlony gel difussion; however, as with fluorescein and rhodamine conjugates, this conjugate should also be titered using tissue reactions.
2. Substrate — This is made immediately before use by dissolving 5 mg of 3,3′-diaminobenzidine tetrahydrochloride (Polysciences®, Warrington, Pa.) in 10 mℓ PBS, pH 7.2. Just before applying the substrate to the section, 100 μℓ of 30% $H_2O_2$ is added with stirring in the dark.
3. Aqueous mounting solution — (Aquamount®, Gurr).

*4. Procedure*

This procedure is very similar to immunofluorescence labeling procedures covered in a previous section. The similarity ends after the peroxidase-coupled IgG secondary (see Reagents section) is applied and incubated with the tissue sections for 40 min. Sections are washed with PBS three times for a total of 10 min, and the substrate solution (see Reagents above) is applied to the slide in a 100-μℓ drop. After a 10-min incubation in darkness at room temperature, the slides are gently washed exhaustively in PBS, covered with Aquamount, and coverslipped. Alternatively the

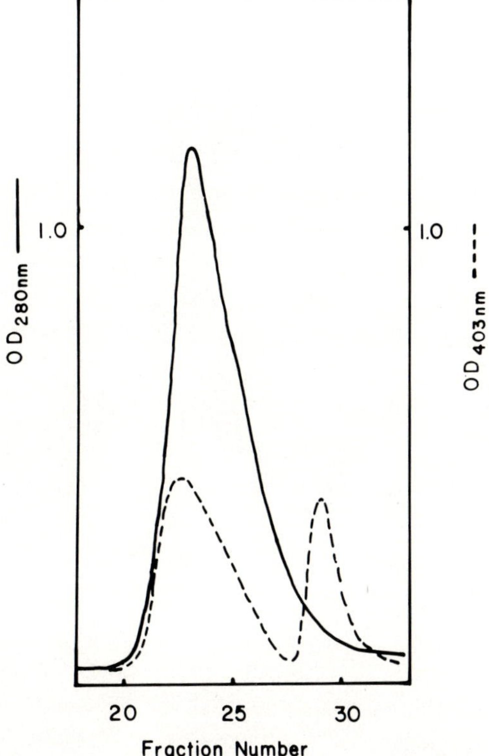


FIGURE 6. (P 150 column) Elution profile of HRPO-IgG conjugate as described in Section II.C. The conjugate elutes in the void volume of the column (tube number 22 to 25), while the unreacted HRPO elutes in the included volume of the column (tube number 27 to 31). The conjugate exhibits considerable absorption at 280 and 403 nm, while unreacted enzyme exhibits absorption primarily at 403 nm.

slides may first be counterstained for 5 min in Hematoxylin and washed exhaustively with water, covered with Aquamount, and coverslipped. They can then be viewed and photographed with a conventional photomicrography apparatus.

*5. Controls*

1. An antibody absorption control in which the primary antibody is absorbed with purified antigen as described previously (see Tissue Immunofluorescence section)
2. A nonimmune serum or IgG control
3. A control for endogenous peroxidase in which the peroxidase-IgG is omitted, but the sections are incubated with substrate

**D. Cell Culture Immunoperoxidase**

*1. Principle*

Refer to previous procedure under Light Microscopic Methods, Reagents (II.A.3).

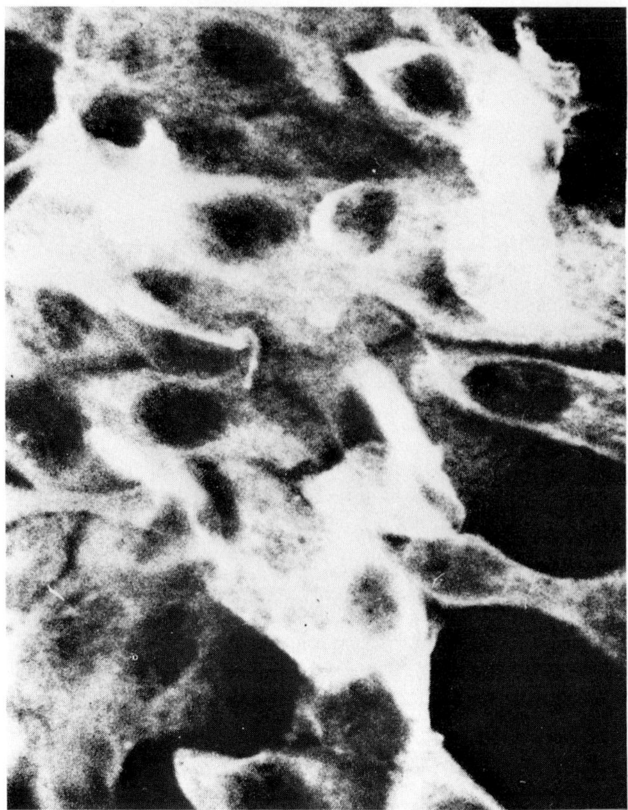

FIGURE 7. Mouse epithelial cells in culture fixed with methanol and stained for type V (AB) collagen. (A) Shows staining of both intra- and extracellular antigen. (B) Illustrates staining of the residual substratum after the cells have been stripped from the culture dish; magnification ×400. (Courtesy of Dr. K. Stenn, unpublished data. With permission.)

*2. Comments*

Immunoperoxidase labeling of cell cultures for light microscopy is basically the same as immunofluorescence labeling and gives the same information without added sensitivity. The advantages of an enzyme tracer are described in (see Section II.C.2 above).

In the method to be described in this section, methanol is used as a fixative. This opens up cell membranes and allows labeling of intra- and extracellular antigens at a light microscopic level (see Figure 7). Unfortunately, ultrastructural detail is not preserved so other fixation methods have to be used for EM work (see Sections III.C. and D). By a variation of the saponin permeabilization method, cells can apparently be fixed and made permeable to peroxidase conjugates in such a way that the same cells can be assessed by both light and electron microscopy.[35]

*3. Procedure*

The procedure is identical to that described for immunofluorescence labeling of such cultures (including methanol fixation) with the following exceptions. Instead of fluorescein-IgG, peroxidase-IgG (see Reagents section above) is applied as the secondary antibody. After a 50-min incubation, the culture dish is washed with PBS

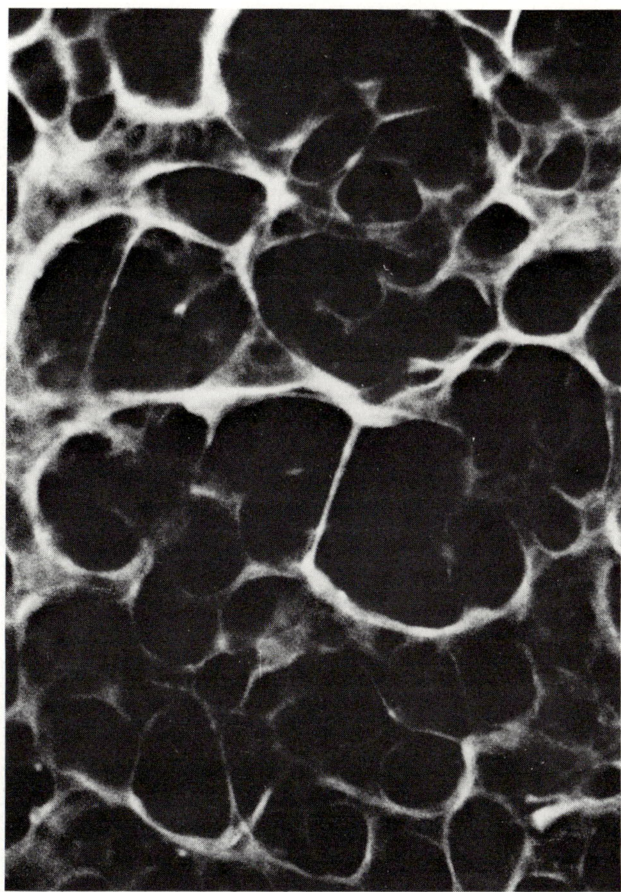

FIGURE 7B

three times, and 100 μℓ of freshly prepared DAB·HCl substrate (see Reagents section above) is pipetted on the cells and incubated for 10 min. This is followed by exhaustive washing and by coverslipping exactly as for tissue sections. Culture dishes or coverslips containing cells can be supported on a slide and viewed and photographed with a conventional light photomicroscope.

*4. Controls*

Refer to previous procedure under Light Microscopic Methods, Reagents (II.A.8).

## III. ELECTRON MICROSCOPIC METHODS

Methods of tissue preparation and fixation that suffice for light microscopic immunochemical work so dramatically alter the ultrastructural appearance of cells that specific organelles are hardly recognizable in the electron microscope. Methods of fixation and embedding developed for conventional electron microscopy, in turn, interfere with immunolabeling of sections. As a result, new methods were developed specifically for immunoelectron microscopy. For comprehensive reviews of these methods readers can refer to chapters by Kraehenbuhl and Jamieson[2] and by Andres et al.[3]

As pointed out by Kraehenbuhl and Jamieson, an ultrastructural localization technique can be classified as either a diffusion localization or a surface localization technique.[2] In diffusion techniques, labeled antibodies are allowed to diffuse into fixed cells or tissue spaces where they can interact with antigens. The tissue or cell block is then washed free of excess labeled antibody, embedded in Epon, and thin-sectioned to display the label which penetrated the tissue and reacted with antigen. In surface labeling methods, the fixed cells or tissues are embedded and thin-sectioned first. Tagged antibodies are then incubated with the thin section and react with the exposed antigen on the surface. Although both methods have advantages and disadvantages which are outlined by Kraehenbuhl and Jamieson,[2] we feel that surface localization techniques have an inherent advantage in that penetration of cell membranes or of dense matrices by tagged antibody is not required. It is clear that ferritin-labeled antibody does not diffuse across cell membranes[36,37] except in the case of active uptake. With surface localization methods, the interior of the endoplasmic reticulum, the midportion of a basement membrane, and the center of a bundle of connective tissue fibers are all equally accessible to antibody. Special diffusion techniques have been designed to overcome the problem of penetration. In this section, we will describe two of them, as well as two surface localization methods, which have been used in connective tissue research.

### A. Tissue: Ultrathin Frozen Sections

This surface localization method is basically that of Tokuyasu,[38] Painter et al.,[39] and Tokuyasu and Singer.[40]

#### *1. Principle*

Lightly fixed tissue is infiltrated with a cryoprotectant such as 2 *M* sucrose. When the tissue is then cooled rapidly to −196°C by plunging it into liquid nitrogen, the remaining cell water in the specimen tends to vitrify instead of forming ice crystals. This is vital since ice crystals are thought to cause most of the morphologic damage to cell structures seen in conventional frozen sections. In addition to minimizing ice damage, the sucrose hardens at these temperatures sufficiently well to provide support for the tissue and allow thin (600 to 800 Å) sections to be cut. The sections can be rapidly warmed and transferred to conventional EM grids. When the sucrose is then washed out, the entire surface of the section is accessible to labeled antibody reagents and antigenicity of the tissue is preserved.

#### *2. Comment*

A number of immunocytochemical studies using this method have now been reported, and it has been repeatedly shown that antigenicity of various proteins and lectins is retained.[19,20,41–45] This might not be too surprising since other cryobiologic work has shown that cryoprotected eukaryotic cells can be cooled with liquid nitrogen and rewarmed with fair survival,[46] implying that vital organelles are not irreversibly damaged. However, for immunocytochemical purposes, preservation of both antigenicity and morphology are of prime importance, and these may not be equated with survival.[47]

The main drawback of the method is morphologic. Some details of cell structure, especially membranes, cannot be seen with the resolution of conventional plastic-embedded sections. On the other hand, details such as myofibrillar fine structure are as well or better seen in ultrathin frozen sections.[48] Tokuyasu has described several methods of improving membrane staining so that it is possible to identify membranes of endoplasmic reticulum, Golgi, mitochondria, and nuclei.[41,49] We have not had enough experience with these refinements to comment on them. Figure 8

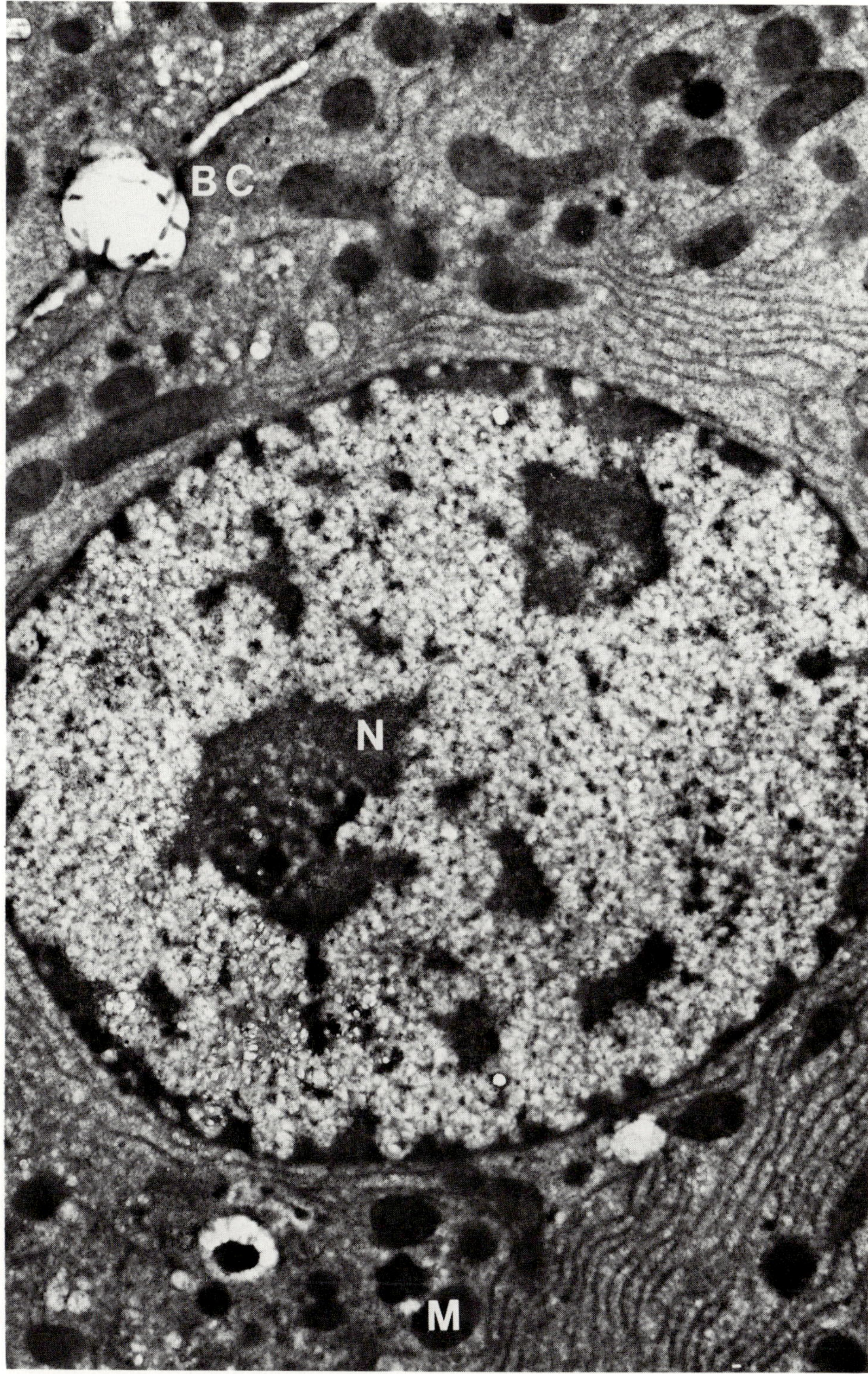


FIGURE 8. Ultrathin frozen section a rat hepatocyte illustrating the preservation of characteristic morphologic features using the technique of Tokuyasu.[40] Tissue is fixed with 1% glutaraldehyde for 10 min and stained with 0.5% uranylacetate. M = mitochondrion, N = nucleus, ER = endoplasmicreticulum, BC = canaliculus; magnification ×75,000.

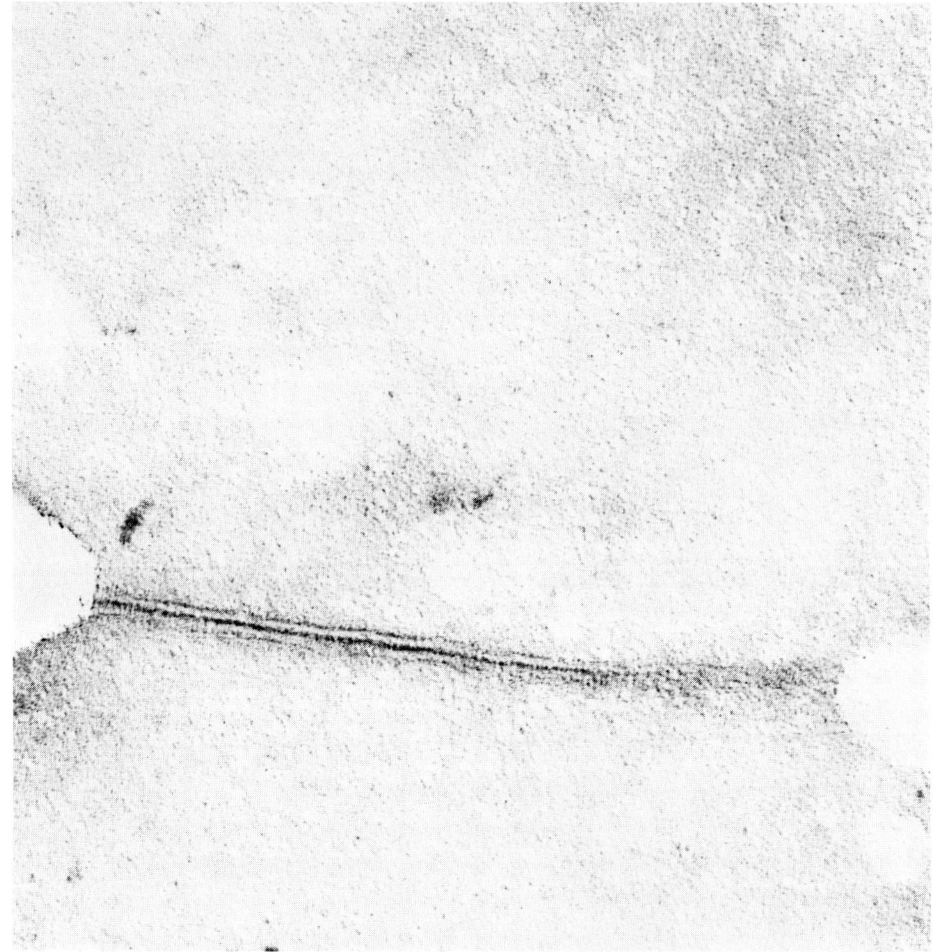

FIGURE 9. Ultrathin frozen section through an hepatocyte gap junction. Although the adjacent plasma membranes are indistinct in this very thin section, the pentalaminar structure of the gap junction is well preserved. Fixation was by immersion in 2% paraformaldehyde and 0.1% glutaraldehyde for 10 min. The section was stained with 0.5% uranyl acetate; magnification $\times$ 156,000.

shows a hepatocyte in frozen section which in most respects is equivalent to a conventional Epon-embedded section. Figure 9 shows a uranyl acetate stained gap junction from a frozen section of a hepatocyte. The trilaminar nature of the junction is clearly seen. On the other hand, adjacent plasma membranes of the hepatocytes are indistinct.

As with frozen sections for immunofluorescence, there is a variety of labeling methods to choose from including one-, two-, or three-stage techniques (see Table 1) and several electron dense markers. We have used ferritin[19,40] and Imposil®[50] — both iron compounds — as markers, but in theory one should be able to use hemocyanin[51] and peroxidase as tracers also.

In general the more stages there are in the immunocytochemical "sandwich", the greater the sensitivity because multivalent intermediates amplify the original signal. The inconvenience of making these reagents and of the extra labeling steps involved is outweighed by the extra sensitivity and, perhaps, lower background achieved.[2]

Another approach used in this laboratory takes advantage of the high binding

**Table 1**
**LABELING ULTRATHIN FROZEN SECTIONS: ONE-, TWO-, AND THREE-STAGE METHODS**

| Number of reagent stages | First stage | Second stage | Third stage |
|---|---|---|---|
| 1 | Primary antibody (e.g., rabbit anti-type I collagen) coupled directly to ferritin | — | — |
| 2 | Primary antibody or primary antibody-biotin | Secondary antibody (e.g., sheep antirabbit) coupled to ferritin or avidin-ferritin | — |
| 3 | Primary antibody | Secondary antibody-biotin | Avidin-ferritin or avidin-Imposil or avidin-horseradish peroxidase |

*Note:* The table lists possible reagent combinations for labeling ultrathin frozen sections. In theory, the ferritin bridge[61] and the peroxidase-antiperoxidase technique[6] could be used also.

affinity of biotin for the glycoprotein avidin.[52,53] Avidin-peroxidase, avidin-iron dextran, and avidin-ferritin can be prepared according to the procedure outlined in this section. For avidin-peroxidase and avidin-iron dextran conjugations, either glycoprotein may be activated by periodate. In the case of avidin-ferritin conjugation, avidin must be activated, as ferritin contains no carbohydrates.

The type of tissue one is studying, however, may restrict one's use of certain reagents. For example, because of the very high affinity of the glycoprotein avidin and the cofactor biotin, ferritin-avidin conjugates will bind to biotin-IgG conjugates applied to frozen sections. However, tissues such as liver which contain endogenous biotin[54] may bind avidin-ferritin nonspecifically. This endogenous biotin cannot be quenched with unlabeled avidin since each avidin molecule has four combining sites for biotin.

Another example of tissue specific restrictions we have encountered is the presence of abundant endogenous ferritin in rat liver, precluding use of this particular marker in studying intracellular antigens in liver cells (see Figure 10). Imposil®, an iron-dextran tracer which has a different shape from naturally occurring storage iron,[50] may be more useful in this situation (see Figure 11).

In this section, we will describe a three-stage method that we have found useful for labeling mouse kidney. For each tissue, choice of the type of label should be governed by consideration of the unique characteristics of the tissue and the chemical properties of each stage of the label.

### 3. *Reagents*

1. Fixative — 0.5 to 1.0% glutaraldehyde (Fisher Scientific Products) in PBS, pH 7.4 — This can be used for either immersion or perfusion fixation.
2. Aldehyde quench — 0.1 *M* glycine–0.05 *M* Tris®, pH 7.4
3. Cryoprotectant — 2 *M* sucrose in PBS, pH 7.4 — This should be kept well stoppered to prevent changes in molarity.
4. Liquid nitrogen — During sectioning this is used at a rate of about 3 ℓ/hr.

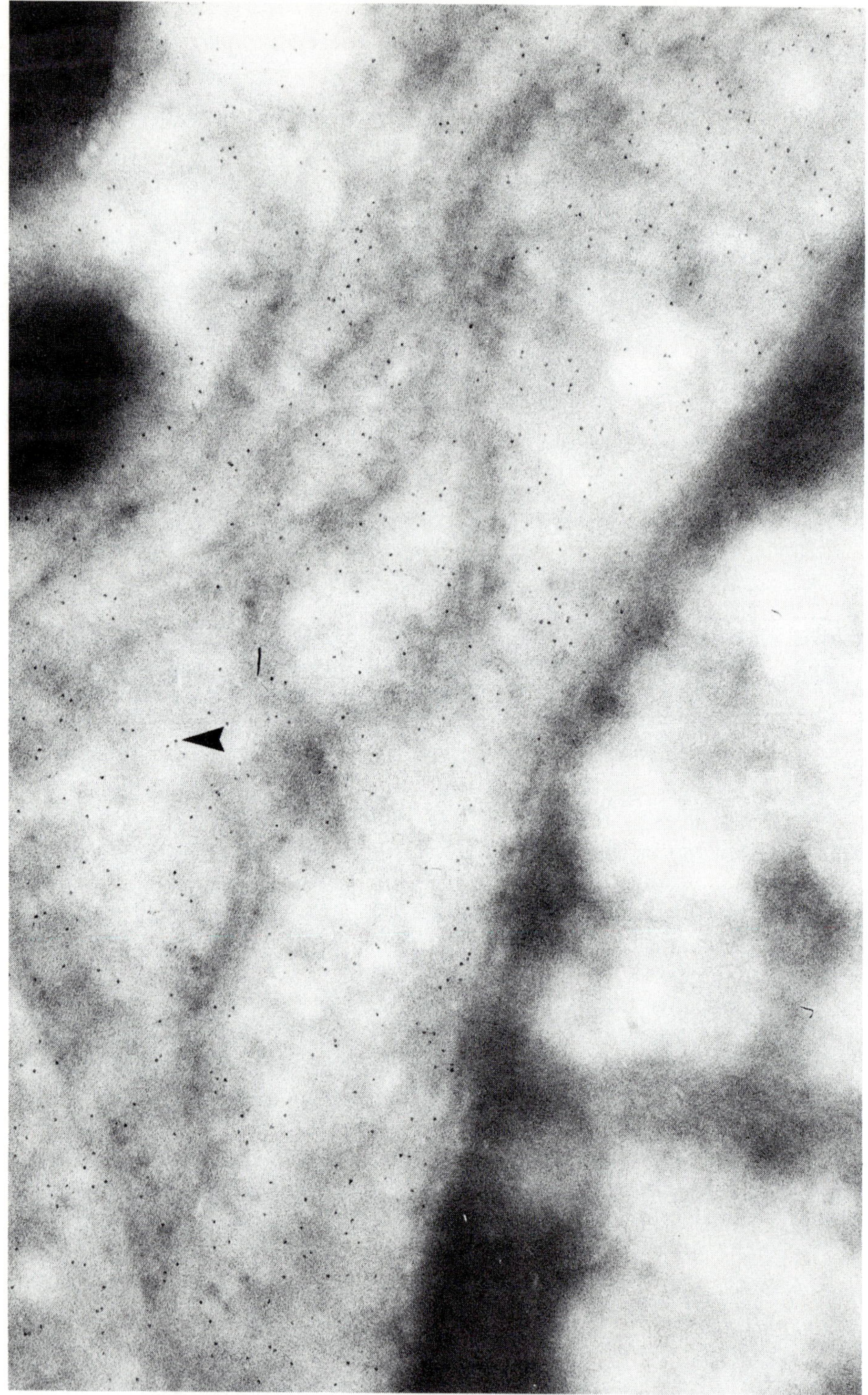

FIGURE 10. Unstained ultrathin frozen section of an hepatocyte from a Sprague-Dawley rat. Numerous electron dense particles can be seen throughout the cytoplasm though not in the adjacent nucleus. These probably represent endogenous ferritin and present a problem in ferritin-labeling studies. N = nucleus; magnification ×156,000.

FIGURE 11. Reticulin fibers from $CCl_4$-treated rat liver labeled with antibody to procollagen III and Imposil®, an iron-dextran conjugate. The sections were labeled as described by Dutton et al.[50] The cigar-shaped Imposil® particles can be readily distinguished from ferritin (see Figure 10); magnification ×156,000.

5. Monochlorodifluoromethane — (Genetron 22®) Allied Chemical Corporation, Specialty Chemical Division, Morristown, N.J.
6. Affinity-purified primary antibody — Antibodies are prepared as described in Chapter II, Volume 1 and diluted to the working dilution with PBS containing 0.1% BSA.
7. Affinity-purified secondary antibody coupled to biotin is prepared by a modification of the method of Becker and Wilchek.[55]
   a. Briefly, 2 mℓ of affinity purified IgG (3.5 mg/mℓ) in PBS is dialyzed into 0.1 *M* sodium bicarbonate, pH 8.0. Biotinyl-*N*-hydroxysuccinimide (9.0 mg) dissolved in 1 mℓ of *N,N*-dimethylformamide (Fisher) is added to the IgG solution. The mixture is stirred for 1 hr at room temperature (22°C) and dialyzed against 0.1 *M* sodium bicarbonate, pH 8.0 at 4°C, followed by dialysis against PBS at 4°C. The biotinyl-IgG is then passed through a 0.22 μ pore size Millipore® filter and stored in 100 μℓ aliquots at −70°C until use. It is diluted to the working dilution with PBS containing 0.1% BSA.
   b. The biotinyl-*N*-hydroxysuccinimide ester is prepared according to the method of Becker and Wilchek.[55] *N*-hydroxysuccinimide (NHS) (2.0 g) is dissolved in 20 mℓ of ethylacetate at 70°C. The NHS is crystallized out of solution by cooling to 4°C. The crystals are collected on Whatman® #1 filter paper using a Buchner® funnel. The crystals are dried over vacuum and ground into a powder and stored in a vacuum dessicator until use. One gram of biotin (Sigma) is dissolved in 60 mℓ *N,N*-dimethylformamide at 50°C in a glass-stoppered bottle. NHS (0.5 g) is added at room temperature with stirring until dissolved. This is followed by the addition of 0.93 g of dicyclohexylcarbodiimide and overnight stirring at 50°C. Any precipitate that forms (dicyclohexyl urea) is filtered out and discarded. The filtrate is evaporated to dryness on a Roto-vap (Buchner) and redissolved in dimethylformamide; as before any precipitated dicyclohexyl urea is filtered out and discarded. The filtrate is evaporated to dryness and redissolved in isopropyl alcohol at 70°C. Any precipitate is taken off by filtration with Whatman® #3 filter paper. The filtrate is cooled to room temperature and the resultant biotinyl-*N*-hydroxysuccinimide crystals form overnight. They are harvested by filtration using Whatman #1 filter paper. If no crystals form, the filtrate may be cooled to 4°C and incubated at this temperature or the filtrate concentrated on a Rotovap® until crystals appear. The biotinyl-*N*-hydroxysuccinimide ester should have a melting point of 186 to 190°C.
8. Avidin-ferritin conjugate — Either of two methods can be used to make this conjugate.
   a. *Glutaraldehyde method of Heitzmann and Richards*[53] — In brief, six times recrystallized horse ferritin (Miles Laboratories®) is crystallized at least one more time from cadmium sulfate ($CdSO_4$) and is freed of ($CdSO_4$) and aggregates as described by Heitzmann and Richards.[53] Avidin® (Sigma Chemical) is dissolved in 0.1 *M* phosphate buffer at 10 mg/mℓ. Approximately equimolar amounts of avidin and ferritin are mixed by adding 0.2 mℓ of avidin solution to 1 mℓ ferritin (20 mg ferritin/mℓ 0.1 *M* phosphate). Conjugation is carried out by adding 0.1 mℓ of a 0.5% glutaraldehyde solution (Electron Microscopy Sciences) to the avidin-ferritin and incubating at 23°C for 1 hr. The reaction is quenched by addition of 2 mℓ of 0.1 *M* $NH_4HCO_3$. This is dialyzed against 0.9% NaCl, passed through a Millipore® filter (0.5 μ pore size) and stored in a sterile vial at 4°C. The working dilution is deter-

mined empirically using tissue known to contain a standard antigen. We use working dilutions of this conjugate in the range of 1 to 20 to 1 to 50.[19]

b. *Periodate-borohydride method* — This method was described by Bayer et al.[56] and is based on the same principal as the IgG-peroxidase coupling method of Nakane and Kawaoi[33], namely mild oxidation of carbohydrate residues of one compound to form aldehydes and subsequent Schiff base formation between the aldehydes and ϵ-amino groups on the other compound. In the case of avidin-ferritin conjugates, avidin must be oxidized as ferritin contains no carbohydrates. In brief, the method of Bayer is as follows:
   1. Dissolve 3 mg avidin in 1 mℓ acetate-buffered saline pH 4.5
   2. Add 0.5 mℓ ferritin (50 mg/mℓ) prepared as described above in method a
   3. Add sodium metaperiodate to 6 m*M*.
   4. Stir for 1 hr on ice.
   5. Stop the reaction by addition of 0.5 mℓ ethylene glycol.
   6. Dialyze the product against acetate-buffered saline for 6 hr at 4°C
   7. Dialyze against borate-buffered saline pH 8.5 overnight
   8. Add 0.15 mℓ of fresh sodium borohydride solution (10 mg/mℓ in 0.01 *M* NaOH) and stir for 1 hr to stabilize the Schiff bases formed.
   9. Dialyze against PBS
   10. Wash 2× by centrifugation at 100,000 × g for 3 hr resuspending the pellet in PBS each time
   11. Resuspend after the last wash in PBS at a ferritin concentration of 7 mg/mℓ.

9. 4% BSA in PBS pH 7.4 — used to pre-treat sections on grids to decrease nonspecific binding of ferritin conjugate to tissue.
10. 0.5% uranyl magnesium acetate stain — Solubilize 50 mg of uranyl magnesium acetate (ICN Pharmaceuticals®, Plainview, N.Y.) in 5 mℓ distilled water. To this add 5 mℓ of 300 m*M* NaCl. The stain is kept at 4°C and passed through a Millipore® filter before use (0.22 μ pore size). The pH should be 4 to 5.

### 4. *Materials*

1. Sorvall® MT2B ultramicrotome (Dupont-Sorvall, Newtown, C.T.)
2. Sorvall® cryokit with Sorvall-Christensen® FTS/LCT-2 (frozen thin sectioner and low temperature control — This liquid nitrogen cooled attachment provides an insulated chamber which can be held at a pre-selected temperature in the range of 0 to −150°C. Specimen and knife (glass knives are sufficient) are held at the same temperature.
3. 17 liter liquid nitrogen Dewar® for cooling the cryochamber
4. 150 mℓ Dewar flask (Balzers, Hudson, N.H.) — This is equipped with a specimen basket (Balzers) to carry the cooled specimen on its chuck to the microtome.
5. 500-mℓ Dewar® flask (Balzers) equipped with a Freon® liquifier for precooling the specimen with liquid Freon® bathed in liquid nitrogen
6. Copper grids for electron microscopy are coated with a 1% Formvar® solution in ethylene dichloride and then given a very light carbon coating which reduces nonspecific binding of ferritin to the Formvar®.
7. Liquid Freon® (Genetron 22®, Allied Chemical, Morristown, N.J.)

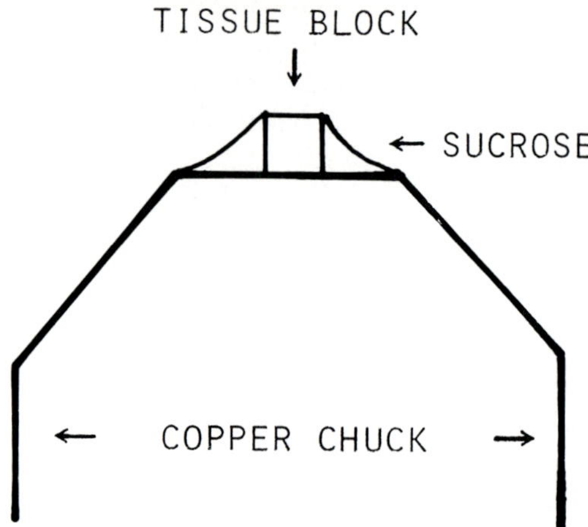


FIGURE 12. Diagram of sucrose-embedded tissue block placed on the end of a copper chuck (here much enlarged) for ultrathin frozen sectioning. (See Section III.A) The sucrose solution has been aspirated from around the tissue block to facilitate cutting.

8. Platinum wire loop — This is made by winding a short piece of 0.005 in diameter platinum wire on itself leaving a 3-mm diameter loop at one end. The other end is fastened to a flamed Pasteur pipette as a handle.
9. Fine forceps for handling EM grids
10. Copper chucks to hold specimens — Chucks can be made from 0.5 cm diameter copper rod stock cut to 1.3 cm length. The specimen holding end is bevelled to provide a 0.3 cm diameter flat surface (see Figure 12).
11. Dissecting microscope for mounting specimen on chuck

*5. Procedure*

***a. Fixation and Cryoprotection***

Fresh tissue is immersed in 1% glutaraldehyde and gently cut into small cubes (less than 1 $mm^3$ in size) with a razor blade. Several dozen cubes are then fixed in 20 mℓ of 1% glutaraldehyde in a vial for a total of 10 to 15 min.

Some organs, such as liver and lung, may require perfusion fixation to obtain good morphology. The fixative is poured off and the cubes are washed with 20 mℓ of PBS twice for a total of 10 min. The glutaraldehyde is quenched by immersion of the tissue in 20 mℓ of 0.1 *M* glycine–0.05 *M* Tris®, pH 7.5 for 15 min. The cubes are then washed with PBS twice, 5 min per wash. The last wash is pipetted off and replaced by 20 mℓ of 2 *M* sucrose in PBS which will serve as a cryoprotectant and embedding medium. The sucrose is allowed to infiltrate the tissue for 1 to 2 hr or overnight, and the tissue blocks are then ready to be frozen and sectioned. The cubes are stable stored in sucrose for at least several weeks at 4°C.

***b. Freezing***

A cube of tissue in a drop of sucrose is transferred in a small pipette (e.g., 100 μℓ) to a glass slide and trimmed with a razor blade to about 0.5 $mm^3$. Under a dissecting microscope, the tissue is then placed on the end of a copper chuck in a drop of sucrose. The size of the specimen and of the sucrose droplet around it are critical for easy sectioning. A specimen which is large will not cut well, and a large sucrose droplet will behave similarly. In addition, if there is too much sucrose, extensive trimming will be necessary before one gets to the tissue. Extra sucrose can be aspirated from around the specimen with a #30 gauge needle and a 1-mℓ syringe. When the specimen protrudes from the surface of the sucrose droplet, enough sucrose has probably been removed (see figure). The chuck is picked up with forceps prechilled in liquid nitrogen, and the end holding the tissue is inverted and plunged for 10 sec in just liquid Freon (−150°C) cooled with liquid nitrogen in the Freon liquifier (see Materials above). After 10 sec the chuck is transferred as rapidly as possible to a Dewar flask of liquid nitrogen (−196°C).

***c. Sectioning***

The chuck can then be transferred with liquid $N_2$ in an aluminum boat to the cryochamber with precooled forceps so that it will not thaw enroute. Finally it is mounted in the cryochamber specimen holder with cold forceps.

Sections are cut with a dry glass knife set at 10° clearance angle with the specimen advance set for approximately 700 to 800 A° thick sections. Good thin sections can be cut manually with the speed selector set at maximum and will come off the block individually or in short ribbons of cellophane-like appearance. During the cutting stroke, the specimen block moves from a colder region of the cryochamber below the glass knife to a relatively warmer region above the knife blade. As it does this, the block expands slightly and thus affects the thickness of the next section. This is not a problem if one finds the optimum rhythm for manual cutting and uses the same rhythm every time. In this medium the usual interference colors are probably not a guide to section thickness.

If sections are curling or adhering to the knife edge, it may be due to an overly warm or humid environment. If occasional sections curl back to the block, they can be flattened down onto the knife with a fine brush so that the block face doesn't crush them on the next pass. The sections may also fly back to the block face because of electrostatic charges but this problem can be eliminated with a Zerostat gun (Ted Pella, Tustin, C.A.). In addition, when trimming the block, it is important not to try to advance the block more than 1 μm at each pass of the knife or the block may be dislodged or fractured. If sectioning is interrupted for a few seconds, the block will expand or contract because of temperature differences in regions of the cryochamber. This can result in the knife either missing the next sections or breaking off a chunk of the block. Consequently it is best to back the knife off 4 to 5 μm and approach the block again every time sectioning is interrupted.

When four or five sections have been collected on the knife they are picked up with a droplet of 2 *M* sucrose in PBS suspended from a platinum wire loop (see Materials above). The droplet is first held in the cryochamber for some seconds until it just freezes (turns opaque) and is then brought close to the sections on the knife blade. Sections will adhere to it, and the loop is lifted out of the chamber and quickly warmed to room temperature. The sections suspended on the underside of the droplet will flatten out on the surface and are transferred to a coated grid by touching the drop to the grid and lifting the loop off. Most of the drop and all of the sections will stay on the grid where they flatten out on the Formvar. Grids can be stored for 1 to 2 days in a moisture chamber at 4°C for subsequent labeling and staining.

***d. Labeling***

The procedure is performed as described by Tokuyasu and Singer.[40] The use of avidin and biotin conjugates was first described by Heitzman and Richards.[53]

A long strip of wide parafilm is tacked to a flat surface to hold drops of reagents.

Reagent drops tend to form beads on the surface, and the grids can be floated face down on the beads for all washing, labeling, and staining steps. All specific reagent drops are 50 μℓ in size. All PBS or water wash drops are about 150 μℓ in size. A pair of fine forceps (Dumont tweezers style #5, Polysciences, Warrington, P.A.) is useful for transferring grids from drop to drop. Eight to fourteen grids, including control grids, is a convenient number for one person to handle. Finally, all labeling reagents including 4% BSA, antibody solutions and ferritin conjugate are centrifuged before use at 20,000 × g for 30 min to reduce aggregates. Each grid containing sections under a sucrose droplet is inverted and floated on the following sequence of reagent drops:

1. PBS × 3 (1 min each)
2. 4% BSA (15 min) to reduce nonspecific labeling by ferritin
3. PBS × 3 (1 min each)
4. Primary rabbit antibody to connective tissue protein (15 min) (see Reagents above)
5. PBS × 12 (1 min each)
6. Biotinized sheep anti-rabbit IgG (15 min) (see Reagents above)
7. PBS × 12 (1 min each).
8. Avidin-ferritin conjugate (15 min) (see Reagents above)
9. PBS × 12 (1 min each)
10. 1% glutaraldehyde in PBS (5 min) as a post fix prior to staining in acidic stain
11. Distilled water × 3 (1 min each)
12. 0.5% uranyl acetate stain (2 min) (see Reagents above)
13. Distilled water × 3 (1 min each)
14. Dry grids by holding each with forceps and touching a piece of filter paper to the edge of the grid
15. Grids examined in the electron microscope (use of a cold trap is advisable) and photographed at low accelerating voltage (60 KV) for higher contrast

6. *Controls*

1. Omission of the first, second, or third-stage reagents on separate grids
2. Substitution of normal rabbit IgG for the first stage antibody
3. An inhibition experiment — In this control the rabbit antibody at twice the working concentration is incubated for 20 min with an equal volume of a freshly prepared solution of the specific antigen in 150 m*M* NaCl–50 m*M* Tris-HCl, pH 7.4. The final inhibitor concentration is usually 100 μg/mℓ. After incubation with inhibitor, the antibody is centrifuged to remove precipitates and used in place of primary antibody.

If background labeling over tissue is high it may help to incubate longer in 4% BSA. Alternatively, unquenched aldehyde groups may be the problem and incubation of the grids on a drop of 0.1 *M* glycine–0.05 *M* Tris®, pH 7.5 for 15 min after Step 1 may solve this. The use of F $(ab')_2$ antibody fragments instead of intact IgG has been advocated because they give lower background labeling in some surface labeling techniques,[2] but we have not had experience with these reagents. If controls indicate nonspecific adherence of the sheep antirabbit IgG, it may help to incubate the sections in nonimmune sheep IgG (0.05 mg/mℓ) for 15 min after Step 1.

## B. Cell Suspension Homogenates

*1. Principle*

Freshly isolated cells are lightly fixed and then disrupted by brief homogenization. Organelles, such as endoplasmic reticulum, Golgi vacuoles, cell nuclei, and mitochondria, are readily identified in the cell fragments, and these structures are then permeable to antibody labeled with ferritin. By appropriate immunolabeling, Epon embedding, and thin sectioning of the cell fragments, the presence or absence of antigen in a particular cell compartment can be ascertained.

*2. Comment*

This diffusion localization method was first described by Olsen and Prockop and associates in a series of papers from 1973 to 1975.[57–59] One of the first applications of immunoelectron microscopy to connective tissue research, this method was used to study intracellular processing of Type I procollagen in chick embryo fibroblasts (see Volume 1, Chapter 7). The technique is designed to overcome the barrier to diffusion of tagged antibody into cells.[59] In their study, although the cells were fragmented, Olsen and Prockop did not observe significant labeling of cell cytoplasm or any other cell compartment than endoplasmic reticulum or Golgi vacuoles. This indicates that the antigen did not leak from protein-synthesizing compartments during homogenization — an important consideration with such a method. The authors showed that leakage from membrane-bound compartments to cytoplasm did occur when fixation was inadequate. On the other hand, prolonged fixation caused a decrease in specific labeling, probably due to more extensive cross-linking of cell proteins which would impede diffusion of antibody reagents. Antigenicity of procollagen was not affected by such prolonged fixation.[59] Another important control done in the course of these studies was incubation of the cell fragments with ferritin conjugated to nonimmune IgG and omission of the washing steps. In this experiment, ferritin was found in all cell organelles except smaller Golgi vesicles, demonstrating that with optimal fixation the labeled antibody could effectively penetrate all organelles with one exception. The simplicity of this method recommends it for studies of intracellular localization of connective tissue components in cell suspensions. One should be able to label by this method the organelles of any type of cell which can be dissociated from its matrix and obtained as a relatively pure population of cells, e.g., smooth muscle cells, endothelial cells, hepatocytes, and various cell lines.

A method of producing ferritin-antibody probes relatively easily was described by the authors and used in the same group of papers.[58] This tracer can be used with any of the immunoelectron microscopic procedures described in this chapter.

The method to be described is taken from References 57, 58, and 59 in which the one-stage label is used to localize intracellular collagen and prolyl hydroxylase in chick embryo fibroblasts.

*3. Reagents*

1. IgG-ferritin conjugate — Affinity-purified rabbit antibody to connective tissue protein is prepared as described in Chapter II, Volume 1. Nonimmune rabbit IgG to be used in control experiments is obtained by ammonium sulfate precipitation as described in Chapter II, Volume 1. Conjugation of both types of IgG with ferritin is carried out as follows. Horse spleen ferritin (six times recrys-

tallized and cadmium free) obtained from Polysciences, Warrington, P.A., is dialyzed into 0.1 *M* phosphate buffer, pH 7.3. For activation, 10 mg of ferritin in 2.0 mℓ of phosphate buffer is mixed with 1 mℓ of 50% glutaraldehyde (Polysciences) with stirring for 30 min at room temperature. The glutaraldehyde is present in 1200-fold molar excess over ferritin to prevent polymer formation. The mixture is centrifuged at 20,000 × g for 10 min to remove a small precipitate, and the supernatant is chromatographed at 4°C on a Sephadex® G-25 (coarse) column, 2.5 × 90 cm, and equilibrated and run with 0.1 *M* sodium phosphate buffer, pH 7.3. This step separates unreacted glutaraldehyde which is retained in the included volume from activated ferritin which elutes in the void. Activated ferritin (approximately 10 mg) in about 20 mℓ of phosphate buffer is immediately mixed at 4°C with 3 mg of affinity-purified rabbit antibody in 1 mℓ of 0.1 *M* phosphate buffer, pH 7.3 for the conjugation step. The mixture is then concentrated to 8 to 9 mℓ with an Amicon PM-30 filter and the reaction is carried out at 4°C for 67 hr. The reaction product is then chromatographed on a 6% agarose column (A-5M, 200 to 400 mesh, (BioRad®)) with dimensions of 0.9 × 60 cm equilibrated and eluted with 0.1 *M* Tris-HCl, pH 7.5 at 4°C. Alternatively, we have used a Sepharose® 6-B (Pharmacia) column with dimensions of 92 × 2.5 cm (see Figure 13). Unreacted aldehydes on the ferritin can be quenched by adding 1 mℓ of 0.1 *M* glycine–0.05 *M* Tris, pH 7.5 to the activated ferritin-IgG mixture 1 hr before it is put on the agarose column. Tris in the running buffer may act as a further quench. Monomeric ferritin coupled to IgG appears as an initial symmetrical peak in the void volume which is monitored by measuring absorbance at 280 nm. Unreacted IgG follows as a separate peak behind the ferritin-IgG. The conjugate is concentrated to about 2 mg of ferritin/mℓ with an Amicon PM-30 filter and stored at 4°C in 0.02% sodium azide. Preparations such as this appear to be stable 6 months or more. The concentration of ferritin in a suspension diluted 1 to 100 can be calculated from the following formula: ferritin (mg/mℓ) $O.D._{440} \times 0.65 \times 100$

2. Collagenase (Sigma®)
3. Modified Krebs medium prepared as described by Dehm and Prockop[60]
4. Trypsin (Grand Island Biological Supply, Grand Island, N.Y.)
5. Eagles minimum essential medium
6. Primary fixative — 1% formaldehyde in 0.06 *M* sodium phosphate buffer, pH 7.3 containing 0.14 *M* sucrose
7. Secondary fixative — 3% glutaraldehyde in 0.6 m*M* sodium phosphate buffer, pH 7.3 containing 0.14 *M* sucrose
8. Post fixative — 1% osmium tetroxide in 60 m*M* sodium phosphate buffer, pH 7.3 containing 0.14 *M* sucrose; embedding medium: Araldite (Taab Laboratories, Reading England)

### 4. *Materials*

Homogenizer, glass, and Teflon® with constant torque motor (Schwaben Präzision, Nördlingen, Germany)

### 5. *Procedure*

1. Isolation of cells — The isolation of chick embryo tendon cells is described in detail[60] and, with slight modifications.[57] In brief, leg tendons are dissected from

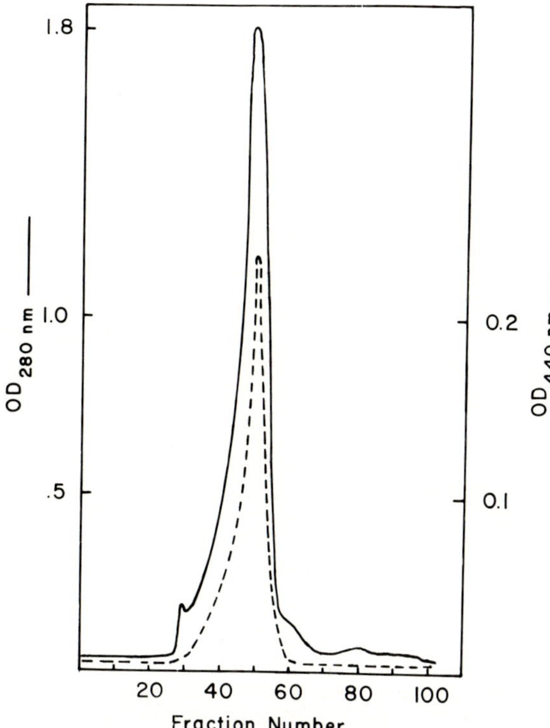


FIGURE 13. Elution profile of ferritin-IgG conjugate from Sepharose® 6-B column as described in Section III.B. Rabbit IgG (3 mg) was coupled to glutaraldehyde-activated ferritin (10 mg) and applied to the column in 5 mℓ of 0.1 *M* phosphate buffer, pH 7.4. The sample is eluted with 150 m*M* Tris buffer, pH 7.4 at 4°C. The fraction size was approximately 5 mℓ. Monomeric ferritin is the large peak eluting in the void volume. Unreacted IgG is found in the small following peak.

17-day-old chicks and digested for 40 min at 37°C in minimum essential medium (2 mℓ/g wet wt tissue) containing collagenase (2.5 mℓ/g wet wt) and trypsin (0.2 mℓ of a 2.5% solution per gram wet wt). Cells are freed of debris by filtration and centrifugation and exposed to soybean trypsin inhibitor. Cells are then washed with medium and an aliquot is taken for counting in a hemocytometer. Yield of cells from one embryo was $11 \times 10^6$ cells; viability of the cells was 84 to 92% (trypan blue exclusion) and $^{14}C$-proline incorporation by these cells upon incubation at 37°C for 2 hr was linear.[60] Immediately after isolation any in vitro metabolic manipulations such as colchicine treatment are performed.

2. Cells are fixed with 1% formaldehyde in phosphate buffer (see Reagents above) for 3 hr at 4°C.
3. Fixed cells are pelleted by centrifugation at 600 × g for 6 min and are resuspended in 0.1 *M* sodium phosphate buffer, pH 7.3 and can be stored overnight at 4°C.
4. Cells are homogenized by 30 to 60 strokes with the Teflon® and glass homogenizer (see Materials above) at 1740 r/min.
5. Cell fragments are collected by centrifugation at 20,000 × g for 10 min.

6. The pellet is incubated with ferritin-conjugated rabbit antibody (see Reagents above) for 24 to 48 hours at 4°C. Homogenate from 50 to 100 × $10^6$ cells is incubated with 100 μg of ferritin coupled to 30 μg of antibody in 100 μl of 0.1 *M* Tris-HCl, pH 7.4 (see Reagents above).
7. Cell fragments are washed twice with 5 mℓ of 0.1 *M* phosphate buffer pH 7.3.
8. Fragments are fixed again with 3% glutaraldehyde (see Reagents above) at 4°C for 1 hr.
9. Post fixation is done with 1% osmium tetroxide (see Reagents above) at 4°C for 1 hr.
10. Labeled homogenate is then block stained with 0.5% uranyl magnesium acetate for 30 min at room temperature, dehydrated in ethanol and embedded in Araldite (or Epon) as for conventional EM blocks. Thin sections are cut and examined in the electron microscope.

#### 6. *Controls*

Two controls appear to be critical:

1. Labeling with nonimmune IgG conjugated to ferritin (see Reagents above)
2. Labeling with nonimmune IgG conjugated to ferritin and omitting the washing step (#7) to ensure that the probe can in fact penetrate the cell fragments

As an additional control one might preincubate the ferritin-IgG conjugate with purified antigen, wash it free of unabsorbed antigen, and then use the conjugate to label fragments. A blocking type experiment in which cell fragments are preincubated with unlabeled specific antibody is feasible, but timeconsuming.

### C. Cell Culture: Saponin Permeabilization Method

#### 1. *Principle*

Lightly fixed cells in plastic dishes can be made permeable to antibody using saponin. In the continuous presence of saponin, antibody will diffuse into cytosol, endoplasmic reticulum, and membrane-bound vesicles, where it can combine with antigen and can be localized with one of a variety of tracers (e.g., ferritin or peroxidase). The fixation and saponin treatment are manipulated to preserve the morphologic appearance of the cell at an ultrastructural level.[61]

#### 2. *Comment*

This diffusion localization method (also called the EGS procedure after the primary reagents used) was developed by Willingham and associates, with the goal of making intracellular antigens of cultured cells accessible to labeled antibodies while preserving ultrastructural morphology reasonably well.[61]

The primary fixatives — ethyldimethylaminopropylcarbodiimide (EDC) and glutaraldehyde were found not to seriously alter antigenicity in tissues studied.[61–64] However this combination of fixatives may make endoplasmic reticulum impermeable to antibody reagents under certain circumstances[65] so that glutaraldehyde alone is sometimes used. A glutaraldehyde concentration which is low enough so that cell membranes can be permeabilized is determined by testing a range of concentration from 0.2 to 1%. The cells are then treated with the detergent saponin to permeabilize them as suggested by Ohtsuki et al.[37] and with nonimmune globulin to reduce nonspecific binding of ferritin. Here the globulin may also function as a quench for the glutaraldehyde. The cells are exposed to primary antibody (also in the presence of detergent) and then to a three-stage "bridge" culminating in a ferritin label.[61] If the antigen is known to be in membrane-bound vesicles, the peroxidase-antiperoxidase technique can be used for labeling since diffusion of reaction product is not significant in this situation.[61] Finally, the cells are fixed again, stained, and embedded in Epon for thin-sectioning.

In published studies, the method has not yet been adapted for localization in intact tissue, but the fixation and permeabilization steps can be used as an alternative to methanol fixation in preparing cell cultures for immunofluorescence.[64] The intracellular localization of fibronectin as well as a number of other proteins have been studied using this technique.[61-65] Because of the reasonably good preservation of morphology, it should be a valuable technique in studying connective tissue proteins in cultured cells at the ultrastructural level. We include a description of the method as used for fibronectin labeling, although we have not used it ourselves.[65]

### 3. *Reagents*

1. Affinity-purified rabbit antibody to connective tissue component prepared as described in Chapter 11, Volume I
2. Affinity-purified sheep antirabbit IgG prepared as described in Chapter 11, Volume I
3. Rabbit antiferritin antiserum purchased from Cappel and affinity purified (e.g., by procedures described for anticollagen antibodies in Chapter 11, Volume I) on ferritin coupled to cyanogen bromide-activated Sepharose® 4-B (Pharmacia)
4. Horse-spleen ferritin (Sigma) at a concentration of 200 μg/mℓ
5. a. Primary fixative (without EDC, used for chick embryo fibroblasts) — 50% glutaraldehyde (Tousimis, Rockville, Maryland) is added to Dulbeccos PBS ($Ca^{2+}$ and $Mg^{2+}$-free) to a final concentration of 0.05%. The solution is made 1 m*M* with respect to EGTA
   b. Secondary fixative (with EDC, used for Swiss 3T3 and NRK cells) — A stock buffer solution is made by mixing equal volumes of 100 m*M* Na phosphate and Dulbeccos ($Ca^{2+}$ and $Mg^{2+}$-free) PBS. Tris® base is added to the buffer to give a final concentration of 1.4% Tris®. The pH is brought to 7.0 with HCl. Powdered EDC (ethyldimethyl-amino-propylcarbodiimide) (Sigma) and 50% glutaraldehyde (Tousimis) are added simultaneously to the buffer to final concentrations varying from 0.05 to 1% and 0.2 to 1% respectively. Exact concentrations depend upon the cell type. As the fixatives are added to the buffer at 23°C a timer is begun and incubation proceeds for 4 min. The pH is adjusted to 7.0 with 1 *N* NaOH at the 3 min mark. At exactly 4 min the fixative is added to the cells (see below)
6. Secondary fixatives — a. 3% glutaraldehyde (Tousimis) in PBS; b. 1.5% osmium tetroxide in 0.1 *M* cacodylate pH 7.0.
7. Dehydrating solutions — 70% ethanol; 100% ethanol
8. Epon 812 mixture
9. Incubation solution — PBS containing saponin (0.05%), nonimmune sheep IgG (5 mg/mℓ) and 1 to 2 m*M* EGTA
10. Stains — a. 1% uranyl acetate in distilled water; b. Lead citrate prepared as described by Reynolds[66]; c. Bismuth subnitrate prepared as described by Ainsworth and Karnovsky[67]

*4. Procedure**

1. Cells in plastic culture dishes are washed twice with PBS (Dulbeccos PBS without $Ca^{2+}$ or $Mg^{2+}$).
2. Cells are fixed in fixative A (see Reagents above) for 5 min at 23°C.
3. Wash with PBS as above.
4. Incubate cells for 30 min at 23°C in incubation buffer (see Reagents above).
5. Antibody incubations (all done in incubation buffer):
   a. Affinity purified rabbit antibody to connective tissue component at approximately 50 μg/mℓ.
   b. Wash with incubation buffer.
   c. Affinity-purified sheep antirabbit IgG at working dilution empirically determined.
   d. Wash with incubation buffer.
   e. Affinity-purified rabbit antiferritin at 100 μg/mℓ
   f. Wash with incubation buffer.
   g. 200 μg/mℓ horse spleen ferritin.
   h. Wash with incubation buffer.
6. Fix cells with 3% glutaraldehyde in PBS for 15 min at 23°C.
7. Fix cells in 1.5% osmium tetroxide in 0.1 *M* cacodylate pH 7.0 for 30 min at 23°C.
8. Dehydrate in 70% ethanol for 3 min and in 4 changes of 100% ethanol for 3 min each.
9. Add 100% Epon directly and equilibrate for 1 to 2 hr.
10. Change to fresh Epon and embed for 2 days at 55°C.
11. The culture dish is broken away from the Epon-embedded cells and small blocks are cut from the Epon.
12. Thin sections are cut and mounted on uncoated copper grids.
13. Sections are stained with 1% aqueous uranyl acetate (30 sec), lead citrate (60 sec), and bismuth subnitrate (30 to 60 min) at 23°C.

*N.B.* If the primary antibody is an antifibronectin antibody, the normal sheep IgG used in all antibody incubations must be rendered free of fibronectin.[65]

*5. Controls*

Controls include dishes labeled with nonimmune rabbit IgG (Step 5A); omission of sheep antirabbit IgG (Step 5C), and omission of rabbit antiferritin (5E).

**D. Cell Culture: Surface Labeling**

*1. Principle*

The surfaces of cells in culture can be labeled using electron-dense, immunochemical markers so that cell surface and matrix components can be localized by electron microscopy. The cell membranes can be sealed by fixation so that only extracellular antigens are accessible and visualized.

* This method was used to label fibronectin in chick embryo fibroblasts.[65] Note that slightly different conditions were used for labeling other proteins in 3T3 and NRK cells.[61,63] Also note that we have changed the source of the antibodies.

*2. Comment*

We have used immunoperoxidase labeling in evaluating the presence of cell-surface associated collagens, in particular Type $AB_2(V)$ collagen in endothelial cell cultures.[24] In order to adequately assess surface labeling, the cell membranes of the cells should be intact and made impermeable to antibody reagents so as not to give false positive results from intracellular staining. We have settled on a 1% glutaraldehyde in PBS fixation for 10 min at room temperature (22°C) and have found this adequate for evaluation of labeling of basal lamina associated collagens (see Figure 5).

*3. Procedure*

The procedure is the same as for labeling cell cultures for light microscopy (see Materials above) except for the fixation step. For surface labeling the cells are fixed in 1% glutaraldehyde in PBS for 10 min at 22°C, followed by a 15 min quench using 0.1 *M* Tris-HCl, pH 7.4 containing 0.1 *M* glycine and 150 m*M* NaCl. This is followed by three brief washes in PBS and incubation with primary antibody. From this point on the sequence is the same as for the previously described immunoperoxidase procedure.

*4. Controls*

Refer to procedure in Section II.A.8.

## ACKNOWLEDGMENT

As far as the authors' work is concerned, it was supported by Grant 5-F-32-AM-05678 (F.J.R.) and 5-F-32-HL-05559 (J.A.M.) from the U.S. Public Health Service and NCI contract 1-CB-84255-37 (H. Furthmayr) from the National Cancer Institute.

## REFERENCES

1. **Coons, A. H., Creech, H. J., and Jones, R. N.,** Immunological properties of an antibody containing a fluorescent group, *Proc. Soc. Exp. Biol. Med.*, 47, 200. 1941.
2. **Kraehenbuhl, J. P. and Jamieson, J. D.,** Localization of intracellular antigens by immunoelectron microscopy, *Int. Rev. Exp. Pathol.*, 13, 1, 1974.
3. **Andres, G. A., Hsu, K. C., and Seegal, B. C.,** Immunologic techniques for the identification of antigens or antibodies by electron microscopy, in *Handbook of Experimental Immunology*, Weir, D. M., Ed., Blackwell Scientific, Oxford, 1978, 37.
4. **Goldman, M.,** *Fluorescent Antibody Methods*, Academic Press, New York, 1968.
5. **Nairn, R. C., Ed.,** *Fluorescent Protein Tracing*, Williams & Wilkens, Baltimore, 1969.
6. **Sternberger, L. C.,** *Immunocytochemistry*, John Wiley & Sons, New York, 1979.
7. **Singer, S. J.,** Preparation of an electron-dense antibody conjugate, *Nature (London)*, 183, 1523, 1959.
8. **Graham, R. C. and Karnovsky, M. J.,** The early stages of absorption of injected horseradish peroxidase in the proximal tubules of the mouse kidney. Ultrastructural cytochemistry by a new technique, *J. Histochem. Cytochem.*, 14, 291, 1966.
9. **Nakane, P. K. and Pierce, G. B.,** Enzyme-labeled antibodies; preparation and application for the localization of antigen, *J. Histochem. Cytochem.*, 14, 929, 1966.
10. **Avrameas, S. and Uriel, J.,** Methode de marquage d'antigens et d'anticorps avec des enzymes et son application en immunodiffusion, *C. R. Acad. Sci. (Paris)*, 262, 2543, 1966.

11. **Wick, G., Furthmayr, H., and Timpl, R.,** Purified antibodies to collagen: an immunofluorescence study of their reaction with tissue collagen, *Int. Arch. Allergy Appl. Immunol.,* 48, 664, 1975.
12. **Timpl, R., Wick, G., Furthmayr, H., Lapière, C. M., and Kühn, K.,** Immunochemical properties of procollagen from dermatosparactic calves, *Eur. J. Biochem.,* 32, 584, 1973.
13. **Olsen, B. R. and Prockop, D. J.,** Ferritin-conjugated antibodies used for labeling of organelles involved in the cellular synthesis and transport of procollagen, *Proc. Natl. Acad. Sci. U.S.A.,* 71, 2033, 1974.
14. **Gay, S., Martin, G. R., Müller, P. K., Timpl, R., and Kühn, K.,** Simultaneous synthesis of types I and III collagen by fibroblasts in culture, *Proc. Natl. Acad. Sci. U.S.A.,* 73, 4037, 1976.
15. **von der Mark, H., von der Mark, K., and Gay, S.,** Study of differential collagen synthesis during development of the chick embryo by immunofluorescence. I. *Dev. Biol.,* 48, 237, 1976.
16. **Timpl, R., Wick, G., and Gay, S.,** Antibodies to distinct types of collagens and procollagens and their application in immunohistology, *J. Immunol. Methods,* 18, 165, 1977.
17. **Bornstein, P. and Ash, J. F.,** Cell surface-associated structural proteins in connective tissue cells, *Proc. Natl. Acad. Sci. U.S.A.,* 74, 2480, 1977.
18. **Ekblom, P., Alitalo, K., Vaheri, A., Timpl, R., and Saxen, L.,** Induction of a basement membrane glycoprotein in embryonic kidney: possible role of laminin in morphogenesis, *Proc. Natl. Acad. Sci. U.S.A.,* 77, 485, 1980.
19. **Roll, F. J., Madri, J. A., Albert, J., and Furthmayr, H.,** Codistribution of collagen types IV and $AB_2$ in basement membranes and mesangium of the kidney. An immuno-ferritin study of ultrathin frozen sections, *J. Cell Biol.,* 85, 597, 1980.
20. **Madri, J. A., Roll, F. J., Furthmayr, H., and Foidart, J.-M.,** Ultrastructural localization of fibronectin and laminin in the basement membranes of the murine kidney, *J. Cell Biol.,* 86, 682, 1980.
21. **Foidart, J.-M., Bere, E. W., Yaar, M., Rennard, S. I., Gullino, M., Martin, G. R., and Katz, S. I.,** Distribution and immunoelectron microscopic localization of laminin, a noncollagenous basement membrane glycoprotein, *Lab. Invest.,* 42, 336, 1980.
22. **Gay, S., Müller, P. K., Lemmen, C., Remberger, K., Matzen, K., and Kühn, K.,** *Klin. Wochenschr.,* 54, 969, 1976.
23. **Alitalo, K., Kurkinen, M., Vaheri, A., Krieg, T., and Timpl, R.,** Extracellular matrix components synthesized by human amniotic fluid epithelial cells in culture, *Cell,* 19, 1053, 1980.
24. **Madri, J. A., Dreyer, B., Pitlick, F. A., and Furthmayr, H.,** The collagenous components of thc subendothelium. Correlation of structure and function, *Lab. Invest.,* 43, 303, 1980.
25. **Johnson, G. D., Holbrow, E. J., and Dorling, J.,** Immunofluorescence and immunoenzyme techniques, in *Handbook of Experimental Immunology,* Weir, D. M., Ed., Blackwell Scientific, Oxford, 1978, 15.
26. **Coons, A. H.,** Histochemistry with labeled antibody, *Int. Rev. Cytol.,* 5, 1, 1956.
27. **Wells, A. F., Miller, C. E., and Nadel, M. V.,** Rapid fluorescein and protein assay method for fluorescent-antibody conjugates, *Appl. Microbiol.,* 14, 271, 1966.
28. **The, T. H. and Feltkamp, T. E. W.,** Conjugation of fluorescein isothiocyanate to antibodies, *Immunology,* 18, 865, 1970.
29. **Emerman, J. T. and Pitelka, D. R.,** Maintenance and induction of morphological differentiation in dissociated mammary epithelium on floating collagen membranes, *In Vitro,* 13, 316, 1977.
30. **Reid, L. M. and Rojkind, M.,** New techniques for culturing differentiated cells: reconstituted basement membrane rafts, *Methods Enzymol.,* 58, 263, 1978.
31. **Peterkovsky, B.,** Effect of ascorbic acid on collagen polypeptide synthesis and proline hydroxylation during growth of cultured fibroblasts, *Arch. Biochem. Biophys.,* 152, 318, 1972.
32. **Siegel, R. C. and Fu, J. C. C.,** Collagen cross-linking. Purification and substrate specificity of lysyl oxidase, *J. Biol. Chem.,* 251, 5779, 1976.
33. **Nakane, P. K. and Kawaoi, A.,** Peroxidase-labeled antibody. A new method of conjugation, *J. Histochem. Cytochem.,* 22, 1084, 1974.
34. **Novikoff, A. B., Novikoff, P. M., Quintana, N., and Davis, C.,** Diffusion artifacts in 3,3′ diaminobenzidine cytochemistry, *J. Histochem. Cytochem.,* 20, 745, 1972.
35. **Hedman, K.,** Intracellular localization of fibronectin using immunoperoxidase cytochemistry in light and electron microscopy, *J. Histochem. Cytochem.,* 28, 1233, 1980.
36. **Morgan, C., Rifkind, R. A., Hsu, K. C., Holden, M., Seegal, B. C., and Rose, H. M.,** The use of ferritin-conjugated antibodies in electron microscopic studies of influenza and vaccina viruses, *Cold Spring Harbor Symp. Quant. Biol.,* 27, 57, 1961.
37. **Ohtsuki, I., Manzi, R. M., Palade, G. E., and Jamieson, J. D.,** Entry of macromolecular tracers into cells fixed with low concentrations of aldehydes, *Biol. Cellulaire,* 31, 119, 1978.
38. **Tokuyasu, K. T.,** A technique for ultracryotomy of cell suspensions and tissues, *J. Cell Biol.,* 57, 551, 1973.

39. **Painter, R. G., Tokuyasu, K. T., and Singer, S. J.,** Immunoferritin localization of intracellular antigens: the use of ultracryotomy, *Proc. Natl. Acad. Sci. U.S.A.*, 70, 1649, 1973.
40. **Tokuyasu, K. T. and Singer, S. J.,** Improved procedures for immunoferritin labeling of ultrathin frozen sections, *J. Cell Biol.*, 71, 894, 1976.
41. **Geuze, J. J., Slot, J. W., and Tokuyasu, K. T.,** Immunocytochemical localization of amylase and chymotrypsinogen in the exocrine pancreatic cell with special attention to the Golgi complex, *J. Cell Biol.*, 82, 697, 1979.
42. **Ziparo, E., Lemay, A., and Marchesi, V. T.,** The distribution of spectrin along the membranes of normal and echinocytic human erythrocytes, *J. Cell Sci.*, 34, 91, 1978.
43. **Griffiths, G. W. and Jockusch, B. M.,** Antibody labeling of thin sections of skeletal muscle with specific antibodies. A comparison of bovine serum albumin (BSA) embedding and ultracryomicrotomy, *J. Histochem. Cytochem.*, 28, 969, 1980.
44. **Beyer, E. C., Tokuyasu, K. T., and Barondes, S. H.,** Localization of an endogenous lectin in chicken liver, intestine and pancreas, *J. Cell Biol.*, 82, 565, 1979.
45. **Cotmore, S. F., Furthmayr, H., and Marchesi, V. T.,** Immunochemical evidence for the transmembrane orientation of glycophorin A. Localization of ferritin-antibody conjugates in intact cells, *J. Mol. Biol.*, 113, 539, 1977.
46. **Rall, W. F., Reid, D. S., and Farrant, J.,** Innocuous biological freezing during warming, *Nature*, 286, 511, 1980.
47. **Plattner, H., Schmitt-Fumian, W. W., and Bachmann, L.,** Cryofixation of single cells by spray-freezing, in *Freeze-Etching Techniques and Applications*, Bennedetti, E. L. and Favard, P., Eds., Societe Francais de Microscopic Electronique, Paris, 1973, 81.
48. **Sjöström, M. and Squire, J. M.,** Fine structure of the A-band in cryo-sections, *J. Mol. Biol.*, 109, 49, 1977.
49. **Tokuyasu, K. T.,** A study of positive staining of ultrathin frozen sections, *J. Ultrastruct. Res.*, 63, 287, 1978.
50. **Dutton, A. H., Tokuyasu, K. T., and Singer, S. J.,** Iron-dextran antibody conjugates: general method for simultaneous staining of two components in high resolution immunoelectron microscopy, *Proc. Natl. Acad. Sci. U.S.A.*, 76, 3392, 1979.
51. **Karnovsky, M. J., Unanue, E. R., and Leventhal, M.,** Ligand-induced movement of lymphocyte membrane macromolecules. II. Mapping of surface moieties, *J. Exp. Med.*, 136, 907, 1972.
52. **Green, N. M.,** Avidin, in *Advances in Protein Chemistry*, Vol. 29, Anfinsen, C. B., Edsall, J. T., and Richards, F. M., Ed., Academic Press, New York, 1975, 85.
53. **Heitzmann, H. and Richards, F. M.,** Use of the avidin-biotin complex for specific staining of biological membranes in electron microscopy, *Proc. Natl. Acad. Sci. U.S.A.*, 71, 3537, 1974.
54. **Moss, J. and Lane, M. D.,** The biotin-dependent enzymes, *Adv. Enzymol.*, 35, 321, 1971.
55. **Becker, J. M. and Wilchek, M.,** Inactivation by avidin of biotin-modified bacteriophage, *Biochim. Biophys. Acta*, 264, 165, 1972.
56. **Bayer, E. A., Skutelsky, E., Wynne, D., and Wilchek, M.,** Preparation of ferritin-avidin conjugates by reductive alkylation for use in electron microscopic cytochemistry, *J. Histochem. Cytochem.*, 24, 933, 1976.
57. **Olsen, B. R., Berg, R. A., Kishida, Y., and Prockop, D. J.,** Collagen synthesis: localization of prolyl hydroxylase in tendon cells detected with ferritin-labeled antibodies, *Science*, 182, 825, 1973.
58. **Kishida, Y., Olsen, B. R., Berg, R. A., and Prockop, D. J.,** Two improved methods for preparing ferritin-protein conjugates for electron microscopy, *J. Cell Biol.*, 64, 331, 1975.
59. **Olsen, B. R., Berg, R. A., Kishida, Y., and Prockop, D. J.,** Further characterization of embryonic tendon fibroblasts and the use of immunoferritin techniques to study collagen biosynthesis, *J. Cell Biol.*, 64, 340, 1975.
60. **Dehm, P. and Prockop, D. J.,** Synthesis and extrusion of collagen by freshly isolated cells from chick embryo tendon, *Biochim. Biophys. Acta*, 240, 358, 1971.
61. **Willingham, M. C., Yamada, S. S., and Pastan, I.,** Ultrastructural antibody localization of $\alpha_2$-macroglobulin in membrane-limited vesicles in cultured cells, *Proc. Natl. Acad. Sci. U.S.A.*, 75, 4359, 1978.
62. **Cabral, F., Willingham, M. C., and Gottesman, M. M.,** Ultrastructural localization to 10 nm filaments of an insoluble 58K protein in cultured fibroblasts, *J. Histochem. Cytochem.*, 28, 653, 1980.
63. **Willingham, M. C., Jay, G., and Pastan, I.,** Localization of the ASV src gene product to the plasma membrane of transformed cells by electron microscopic immunocytochemistry, *Cell*, 18, 125, 1979.
64. **Willingham, M. C., Yamada, S. S., and Pastan, I.,** Ultrastructural localization of tubulin in cultured fibroblasts, *J. Histochem. Cytochem.*, 28, 453, 1980.

65. **Yamada, S. S., Yamada, K. M., and Willingham, M. C.,** Intracellular localization of fibronectin by immunoelectron microscopy, *J. Histochem. Cytochem.*, 28, 953, 1980.
66. **Reynolds, E. S.,** The use of lead citrate at high pH as an electron-opaque stain in electron microscopy, *J. Cell Biol.*, 17, 208, 1963.
67. **Ainsworth, S. K. and Karnovsky, M. J.,** An ultrastructural staining method for enhancing the size and electron opacity of ferritin in thin sections, *J. Histochem. Cytochem.*, 20, 225, 1972.

Chapter 4

# THE USE OF ANTIBODIES TO CONNECTIVE TISSUE PROTEINS IN STUDIES ON THEIR LOCALIZATION IN TISSUES

**Heinz Furthmayr and Klaus von der Mark**

## TABLE OF CONTENTS

# I. INTRODUCTION

The immunochemical characterization of antibodies to collagens, procollagens, and the connective tissue glycoproteins fibronectin, laminin, and chondronectin, has provided very useful tools to study questions of biological interest. Connective tissue as the broadly defined matrix fills in the extracellular space in essentially all organs in addition to forming some specialized structures of great stability. Besides this structural role to build a framework to which cells can attach, it almost certainly influences cell behavior. It is extremely important as a matrix on which cells migrate, that affects developmental processes or that contributes to maintaining cellular differentiation in organs.

There is a multiplicity of collagens, connective tissue glycoproteins, as well as proteoglycans, which form connective tissue. Since different cells synthesize these molecules or combinations of them, it seems appropriate to ask the question of specificity. Is there specificity to particular connective tissue elements in differentially influencing cell behavior? In other words, which combination of components and which macromolecular organization is required to support any one of the particular cellular functions of cell adhesion, cell migration, cell proliferation, or differentiation?

Antibodies to connective tissue elements have been applied to identify metabolically labeled products of cells in tissue culture, to serve as diagnostic tools in quantitative radioimmunoassays for experimental and clinical studies, and to study the distribution in organs and specialized tissue structures of connective tissue macromolecules in vivo. The information gathered from the latter studies combined with the insight provided from in vitro experiments may bring us closer to a critical understanding of the complex cellular functions mentioned above.

In this article, we will restrict ourselves to reviewing the information obtained by applying antibodies to tissue sections in immunohistological studies at the light and electron microscope level.

# II. CRITICAL REMARKS TO THE METHODOLOGY OF IMMUNOHISTOLOGICAL TECHNIQUES

The most widely used techniques for studying the localization and distribution of connective tissue macromolecules include the indirect immunofluorescence and immunoperoxidase technique for light microscopy and the immunoferritin and immunoperoxidase technique at the ultrastructural level. The advantages, disadvantages, and various limitations ("pitfalls") of these techniques have been described in many articles (referenced in Reference 1), and we will limit ourselves here to a critical evaluation of the work on connective tissue only.

Immunohistological and cytochemical methods are quite useful for connective tissue research, but as a number of discrepant results in the past and more recently show, reliable results and interpretations depend on critical application and appropriate control experiments. However, even with careful use of these methods, ambiguities cannot always be completely avoided. Since all of the methods have limitations of one sort or another, the solution to resolve discrepancies may not always lie in applying a combination of two or more morphological techniques, but rather to seek an answer by independent technology.

Some of the pitfalls and discrepancies observed in the past should serve to illustrate the basis for these remarks.

One of the first studies by immunofluorescence used antibodies to collagen (presumably Type I) isolated from the skin.[2] The raw antiserum "stained" interstitial connective tissue and glomeruli in the kidney, and it was concluded that glomerular basement membranes contain collagen. In a later, more carefully controlled study, it was shown unambiguously, that kidney glomeruli do not react with antibodies to Type I collagen.[3] Although the unpurified, raw antiserum showed the same staining of interstitium and glomeruli as in the study mentioned before, the antibodies isolated by affinity chromatography on Type I collagen reacted only with the interstitium, but not with glomerular basement membranes. The antibodies which stain the latter structures could, however, be recovered in the nonbound protein fraction after passage through the immunoadsorbens. These do not react with collagen, but rather with other unidentified material. The lack of cross-reactivity of antibodies to Type I collagen with the later discovered basement membrane collagens is well documented now.[4,5] This example describes well the first pitfall of "false positives" due to poorly characterized antibodies. Using raw antisera can be satisfactory for some immunological studies and particularly when the antigens used in immunochemical reactions are well characterized, chemically pure, and when the reaction can be properly quantified. When complex antigen mixtures such as present in tissue sections are analyzed, the antibody reagent has to be purified and well characterized because of the risks of detecting unrelated material. Such unrelated antigens potentially contaminate the preparation used for raising the antiserum, and antibodies are obtained to the antigen proper as well as the contaminant(s). This is particularly true for collagen antisera for several reasons:

1. Collagens are ubiquitously present in many or all organs and so may be a contaminating immunogenic substance—Isolation of collagen as the immunogen from one site does not guarantee that the reactivity of the raised antibodies in the antiserum for a different tissue is collagen specific
2. Some of the collagens are weak immunogens, and several injections of the antigen are required to get reasonable antibody titers — This may lead to amplification of the immune response to minute amounts of chemically virtually undetectable impurities — Antibodies to impurities such as serum proteins in fact can be readily identified in many collagen antisera
3. The different types of collagens are related chemically, and cross-reactivity has been observed of antibodies raised to one type with other types — Cross-reactivity with related antigens thus can cause confusion in studies on tissue if unrecognized

All these factors provide sufficient rationale for extreme precautions: to use affinity chromatography with heterologous collagen types (to avoid cross-reactions) and final adsorption and elution from the homologous affinity system (to recover highly specific antibodies). In spite of the precautions, one has to be aware of the fact that adsorption with known collagen types does not exclude possible cross-reactions with as yet unknown collagens or that the levels of contaminants are high enough to make the conventional approach of antibody purification by immunoadsorption less useful. These antibody purification schemes are based on the often valid assumption that binding of antibodies of mixed specificities to an affinity system containing the same antigen (and impurities) as the preparation used for immunization will result in a considerable decrease (to undetectable levels) of antibodies to contaminants in the bound and eluted antibody fraction. This assumption is reasonable since binding is dependent on stoichiometric relationships and contamination with noncollagenous proteins are usually small. For example, positive immunofluorescence was observed in the subendothelium of blood vessels[6] and in the early embryo[7] with antibodies to Type III collagen in some but not all laboratories.[8,9] These conflicting results could be due to cross-reactive antibodies present in antisera specific for Type III collagen, but they could also be due to variation in the recognition of antigenic determinants by different antisera or they are due to other problems (see below).

The alternative possibility to raise highly specific antibodies has become feasible through development of monoclonal antibodies by somatic cell hybridization (see Chapter 12, in Volume 1). The drawback of such antibodies is their extremely high specificity. Although conventional antibodies raised in rabbits can react with the same type of collagen from a number of species, monoclonal antibodies rarely do, and presumably such antibodies have to be developed for every experimental system under study.[10]

Another "pitfall" is illustrated in studies with Type I collagen antibodies in normal and pathological human liver.[11] In this study, antibodies to Type I collagen did not stain sections, although in humans this type of collagen is a prominent component of connective tissue.

The lack of reactivity on tissue sections of antibodies to connective tissue proteins creates thus new "pitfalls." Antibodies to Type II collagen do not stain the matrix of hyaline cartilage unless the tissue section is treated with hyaluronidase.[12] Hyaluronidase treatment was also required to detect intracellular fluorescence staining of chondrocytes with antibodies to the core protein of chondroitin sulfate[13] or to stain early mouse embryos with antibodies to Type IV collagen or to laminin.[7] In studies on skin, antibodies to Type III collagen stained the upper dermis intensely, but virtually no staining was observed in the deeper layers.[14] Biochemical analysis[15] and studies on fibroblasts from the different layers[16] did not reveal such a gradient of production and deposition of Type III collagen in the skin.

Thus, accessibility of these proteins in the tissues will influence the staining result, not only to give positive or negative reactivity, but also to tempt investigators to make quantitative judgments based on fluorescence intensity. Although the quantum yield can be rather accurately determined and may in fact reflect in some systems the amount of a given antigen in a cell or in tissues after careful standardization,[17] some of the examples given above should serve to prevent creating the illusion that quantitation of immunofluorescence images necessarily provides more "solid" data. Only in conjunction with additional information will this be useful. One of the more striking examples of the lack of correlation of fluorescence intensity and amount of collagen actually present can be found in articles on fibroblasts.[18,19] Although by immunofluorescence or immunoperoxidase no difference can be seen between Type I and III antibody reactivity in fibroblasts, the cells produce much more Type I than Type III collagen, and the ratio of the two is changing, depending on the number of cell generations.

Other problems, such as nonspecific reactions of fluorescein-labeled secondary antibodies, autofluorescence of tissue components, the need for proper controls, the titration of primary, and secondary antibody for "optimal" fluorescence, have been elaborated in other texts, and they will not be further dealt with here.

One of the major limitations of immunohistological techniques using the light microscope is their low resolution. Because of the nature of the extracellular matrix and its location around and between cells on the one hand and its organization to more elaborate structures on the other hand, these techniques often allow qualitatively to assess whether a particular protein is present in a given tissue. In some instances, e.g., cartilage or bone, this is quite sufficient information. In other instances, however, the resolution is not good enough to decide whether a particular antigen is associated with a basement membrane or the cell adhered to it or whether antibody staining occurs right at the interphase of the two structures. For this reason, ultrastructural techniques are desirable which would combine high resolution of structures in optimally prepared tissue with immunocytochemical probes. In general, this requires stabilization of tissue by fixation of one kind or another to preserve cell and tissue architecture. Since optimal fixation for preservation of the ultrastructural appearance does not necessarily coincide with preservation of antigenic reactivity of different cellular and matrix constituents, a large number of variations of the basic techniques described in Chapter 3 in Volume II have been introduced. These variations are in most instances determined empirically and are influenced by the need to find a workable range of conditions for the two competing requirements — fixation and preservation of antigenicity. This range indeed can sometimes be very narrow.

Novikoff et al.,[20] in their article on diffusion artifacts of the peroxidase cytochemical techniques, state that "cytochemical reagents may never be free of limitations", and they urge investigators to use caution when interpreting cytochemical observations. In spite of appropriate controls, the data will be influenced by a number of variables which cannot always be easily controlled. The advantages and disadvantages of the two principal techniques — diffusion localization and surface localization — have been outlined in other texts.[1,21] Both techniques have been applied to connective tissue research, and some controversial results have emerged which will be discussed in more detail below.

The major thrust for immunocytochemistry at the ultrastructural levels is derived from its great potential to visualize individual components in their original relationship and to reconstruct the original tissue organization of biochemically isolated molecules. This information cannot be easily obtained by other approaches. It could eventually allow correlation of tissue structure with functional implications. With newly isolated molecules, immunocytochemical staining, however, invades a frontier of knowledge that permits data only to be accepted provisionally. In the absence of other information, only consistency of results obtained with these techniques and in different laboratories allows findings to become established. Only the most crucial parameters will be outlined here, and as more experience is gained and different methods applied, immunocytochemistry may well become a crucial element in understanding the macromolecular organization of connective tissue.

The sensitivity of the techniques is steadily increasing, and some of the multistage procedures such as the peroxidase-antiperoxidase technique[22] are among the most sensitive methods for detection of antigen. Since the concentration of the antigen of a particular tissue is usually unknown, the sensitivity can only be deduced by comparison with a tissue accessible to biochemical methodology. This has rarely been done in other systems[22] and has not been done for connective tissue. Differences in structural organization could easily influence the results, and thus such data are difficult to come by. Quantitation of "staining" is not easily possible, and the presentation of data in addition is dependent on selection of "representative" images by the investigator. Since "nonspecific" labeling is inherent in all these techniques, a sufficient signal to noise ratio is required which can vary considerably. The significance of small numbers of ferritin particles or low amounts of the oxidation product of the peroxidase reaction thus is not always clear.

Because of the sensitivity of the immunoelectron microscopic methods, it becomes crucial that the specificity of the antibodies has carefully been established with the most sensitive immunochemical methods. Although the idea of using the most specific reagents—monoclonal antibodies—appears attractive, it will still have to be shown whether the assurance of great specificity is not outweighed by the disadvantage of low numbers of reactive sites (1/mol) in the tissue.

Since the methods of fixation and plastic embedding developed for conventional electron microscopy prevent labeling of tissue sections with antibodies to most collagens and connective tissue glycoproteins, methods have to be used which preserve antigenicity. One such method is the surface labeling of ultrathin frozen sections of lightly fixed tissue, and the second method which has been applied is preembedding staining of fixed tissue by diffusion of reagents into small tissue blocks followed by plastic embedding and sectioning. The second method suffers in general from penetration problems. Diffusion of the reagents into the tissue occurs only up to a certain depth and nonuniformity of staining can often be observed. Diffusion of excess reagent out of the tissue block leads to further possibilities of nonspecific reactions. It is said that the size of the reagent or conjugate is not necessarily the limiting factor.[22] Tissue preservation is much better, however. Staining of frozen ultrathin sections with immunohistochemical reagents is presumably not different from staining of frozen sections for immunofluorescence light microscopy. In order to obtain good sections and to preserve antigenicity, one has to compromise by using less stringent fixation, and further development of the method will be required to achieve the desired preservation and resolution of tissue structure. Damage of the cells due to ice crystal formation and the difficulty to get good sections from every tissue are some of the present drawbacks.

In summary one can conclude that up until now the full repertoire of immunohistochemical technology has not been fully explored. This is particularly true for ultrastructural studies. With the availability of well characterized antibody reagents now, it is expected that the controversial results will be resolved and new data accumulated. Such data are clearly needed in order to understand the biological complexity and function of connective tissue and its influence on cells.

## III. LOCALIZATION OF COLLAGENS IN DIFFERENT ORGANS

### A. General Distribution of Collagens

Because of its simplicity and sensitivity, the indirect immunofluorescence technique has been most widely used for localizing collagens in tissues and cells. As discussed above, affinity-purified antibodies are essential for obtaining reliable results. With the exception of perhaps only a few tissues (such as the lens capsule), most tissues and organs contain two or more collagen types. In organs, such as skin, lung, placenta, or liver, and tissues, such as cartilage or bone, biochemical analyses have confirmed the presence of several collagens. Obviously, biochemical analysis cannot distinguish the source or exact locale of a small amount of a particular type of collagen in any tissue containing for instance blood vessels, glands, or other structures in addition to organ specific cells.

Although at the light microscopic level various collagen types frequently are codistributed, e.g., Types I and III collagen in reticular fibers and Types I and V in the embryonic cornea,[23] some tissues, such as bone, cartilage, or tendon, contain predominantly a single type, and such tissues can be used to evaluate the specificity of antisera. Based on previous experience and data reproduced in several laboratories, more complex organs such as the kidney can serve the same purpose, since only particular types of collagens are associated with glomeruli, blood vessels, basement membranes, etc.

The data to date show that immunofluorescence technology is quite useful to screen tissues for collagens and to eventually assign a particular localization or association with recognizable structures. It was possible to detect the five types of collagens and procollagens of Type I and III in a variety of tissues. Table 1 summarizes the present information. In using this data, the reader should be aware of the fact that not in all cases is confirmation available and thus should approach it with a critical attitude. In addition, some or most of the studies did not have available the full set of antibodies to all the collagen types known now, and additional work will still be required to obtain complete information. Many of these studies also do not address the possibility of species-, age-, or unknown disease-related differences adequately. In general, Type I collagen was identified with biochemical and immunological methods in nearly all tissues with few exceptions, e.g., hyaline cartilage, vitreous humor, and basement membranes. It is in most tissues associated with Type III collagen, e.g., in reticular fibers in muscle[30] or skin.[24,25,29] However, bone matrix,[80] embryonic tendon,[81] or cornea[23,82,83,116] contain Types I and V collagen, but no Type III collagen.

Type II collagen is mostly found in hyaline and elastic cartilage,[84,85] but it is also present in embryonic[86] and in sturgeon notochord,[87] vitreous humor,[88] and in the primary corneal stroma of the chick embryo.[89] Type IV collagen has been located biochemically and with antisera against different forms and fragments of Type IV collagen consistently in all basement membranes,[26,39] but also in the inner cell mass of a mouse blastocyst.[7]

Type V collagen is associated with striated and smooth muscle cells and with basement membranes,[43,76] but is also found in the interstitium of many organs even if basement membranes or a basal lamina morphologically are not distinguished.

### B. Special Tissues

#### *1. Connective Tissue Proper*

The major collagenous constituents of soft connective tissues are Types I and III collagen (see also sections on liver and lung). Type I and III collagen were localized by immunofluorescence in the periosteum[33] and perichondrium[142] and in organ capsules and reticular fibers of the liver[46] or muscle,[30] spleen,[29] and adipose tissue (see Figure 1a and b). In earlier studies, the failure to stain reticular fibers with antibodies to Type I collagen[11,29,46] may have been due to masking of antigenic sites or low titers of antibodies to Type I collagen. The classical histological concept which distinguishes between collagenous and reticular fibers on the basis of silver staining thus did not find a structural correlate related to differences in collagen type. Organ specific differences of connective tissue fibers may exist, however, These could be due to differing dimensions of collagen fibrils, different ratios of Type I to III collagens, or the presence of other structural macromolecules, e.g., fibronectin[91] or elastin.

#### *2. Skin*

Biochemical studies on the content of Types I and III collagen in human skin, based on the quantitative comparison of CNBr-peptides, indicated that Type III collagen predominates in early fetal skin, but represents only 20 to 25% after birth and in later life.[92,93] Similarly, considerably more Type III collagen than Type I collagen synthesis was observed in guinea pig skin during fetal development, while in the skin of newborn animals, Type III collagen accounted only for about 10% of the total collagen.[94]

Immunofluorescence studies on the distribution of Types I and III collagen in skin

**Table 1**
**IMMUNOHISTOLOGICAL LOCALIZATION OF COLLAGENS AND PROCOLLAGENS IN TISSUES**

| Tissues or organ | Collagens detected[a] | Technique[b] | Ref. |
|---|---|---|---|
| | Normal Mature Tissues | | |
| Skin | I, III, IV, pI, pIII | IF | 24–26 |
| | III | EM | 27 |
| | IV | IP | 28 |
| Tendon | I, III, V | IF | 19, 29, 30 |
| Bone | I, II | IF | 31–33 |
| Hyaline cartilage | I, II, III | IF | 31, 32, 34, 35 |
| | V | IF | 36 |
| Intervertebral disk | I, II | IF, IP | 35, 37 |
| Blood vessels | I, III | IF, IP | 6, 8, 38 |
| | IV, V | IF, IP | 8, 39 |
| Trachea | I, II, III, IV | IF | 40 |
| Lung | IV, V | IF | 39, 41 |
| Kidney | I, IV, pI, pIII | IF | 3, 39, 42 |
| | IV, V | IF, EM | 42, 43 |
| Liver | I, III, IV, V, pI, pIII | IF | 45–49 |
| | I, III, IV, V | IF, IP | 50 |
| Spleen | I, III | IF | 29 |
| Muscle | I, III, V | IF | 30, 51 |
| | pI, pIII, IV | IF | 52 |
| Intestine | III, IV, V | IF | 53 |
| Nerve | IV, V | IF | 54 |
| Mummified skin | I, III | IF | 55 |
| | Developing Tissues | | |
| Early embryo, mouse | I, III, IV, V, pIII | IF | 7, 9 |
| | I, IV | IP | 56 |
| Early embryo, chick | I, II | IF | 12, 57 |
| Eye, chick | I, II, pIII | IF | 58 |
| | V | IF | 23 |
| | IV | IF | 59 |
| Bone formation, mouse | I, II | IF | 60 |
| Bone formation, chick | I, II | IF | 34, 57 |
| None formation, implant | I, II, III, V | IF | 61, 62 |
| Kidney tubulogenesis | I, IV, pIII | IF | 63 |
| Tooth formation | I, III, IV, pIII | IF | 64, 65 |
| Muscle formation | I, III, V | IF | 66 |
| Amniotic membrane | I, IV, pIII | IF | 67 |
| | Pathological Tissues | | |
| Skin, scleroderma | I, III, IV, pI, pIII | IF | 68, 69 |
| Skin, dermatosparaxis | I, pI | IF, EM | 70 |
| Skin, various disorders | I, III, IV, pI, pIII | IF | 24, 71, 72 |
| Liver fibrosis, human | I, III, V, pI, pIII | IF | 45, 47, 49, 73 |
| | IV | IF | 48 |
| | I, III, IV, V | IF, IP | 50 |
| Liver fibrosis, rat | I, III, pIII | IF | 74 |
| Lung fibrosis | I, II, III, IV, V, pI, pIII | IF | 75, 76 |
| Scar and granulation tissue | I, III, pIII | IF | 77, 78 |
| Joint and bone lesions | I, II, III | IF | 31, 33, 79 |
| Artherosclerotic plaques | I, II, III | IF | 6, 53 |

[a] Collagens Type I, II, III, IV, or V and procollagens (pN-collagens) Type I (pI) and III (pIII).
[b] IF = indirect immunofluorescence; IP = immunoperoxidase reaction (light and/or electron microscope); EM = electron microscopy after ferritin labeling.

suggest an even distribution of Type I collagen throughout all levels of the dermis,[25,53] while Type III collagen is apparently more prominent in the adventitial dermis.[25,29] These observations are, however, at variance with a study by Epstein and Munderloh,[15] who found an even distribution of Types I and III collagen in all levels of the dermis with biochemical techniques.

Ultrastructural studies by Fleischmajer et al.[27] with ferritin-labeled antibodies to Types I and III collagen have shown that Type III collagen assembles to fibrils of about 250 Å diameter, while the thicker fibrils of the skin (800 to 100 Å diameter) appear to consist of Type I collagen.

Antibodies to the aminoterminal extension of Type I procollagen labeled predominantly the stratum papillare of normal skin,[3,95] while in skin of dermatosparactic sheep and cattle, *p*-collagen is evenly distributed throughout the dermis.[70] The significance of these findings is unknown. Type IV collagen is located in the basement membrane of the dermal-epidermal junction, around hair follicles and sweat glands, and in capillaries in mouse skin[28,96,97] and in human skin.[72]

*3. Tendon*

The major collagen found in tendons is Type I, where it forms thick bundles of fibers. The sheath of reticular fibers which are wrapped around individual fiber bundles, the endotendineum, consists, however, of Type III collagen.[19,29,30] Biochemical studies on the other hand describe the major collagen as Type I,[98] and Jimenez et al.[81] found in addition about 8% acid-soluble Type V collagen in lathyritic chick embryos. Chung et al.,[99] however, did not detect this type in bovine achilles tendon, nor has Type III been isolated.

*4. Bone*

In adult bone, Type I collagen is seen by immunofluorescence (see Figure 2) only after demineralization with EDTA.[31,32,34] It is located in the trabeculae of chick and human cranial bone[12,53] and in periosteal and endochondral bone (see Figure 2C) of the growing chick tibia[34] and of mouse bone,[60] as well as in ectopically formed bone matrix after s.c. implantation of demineralized bone powder.[61] Trabeculae of spongy bone in growing long bones contain a core of Type II collagen which is located in the remaining calcified cartilage after cartilage-bone metamorphosis.[32,33]

Type III collagen is normally absent from the bone matrix. However, this collagen is found in association with the reticular endothelium of bone trabeculae, especially of growing bone, and in holes of the compact bone of patients affected with osteogenesis imperfecta. It is not present in osteoblastomas or osteosarcomas[33] which contain predominantly Type I collagen.

*5. Cartilage*

Extraction of the xyphoid cartilage from lathyritic chicken with neutral salt solutions and acid led to the first demonstration of the genetic heterogeneity of collagen.[84]

Besides Type I collagen, a cartilage-specific collagen, called Type II, was described which gave rise to a different peptide pattern after cleavage with CNBr.[90] Trelstad et al.[101] concluded that the cartilage matrix itself consisted only of Type II collagen, while the Type I collagen found in acid extracts of cartilage was derived from remainders of the perichondrium. This conclusion was later confirmed by immunofluorescence analysis of chick sternal cartilage with antibodies specific for Types I and II collagen.[12,34]

In chick cartilage, the epiphyseal, growth, columnar, and hypertrophic regions contain only Type II collagen. However, Type I is found in chicken also on the

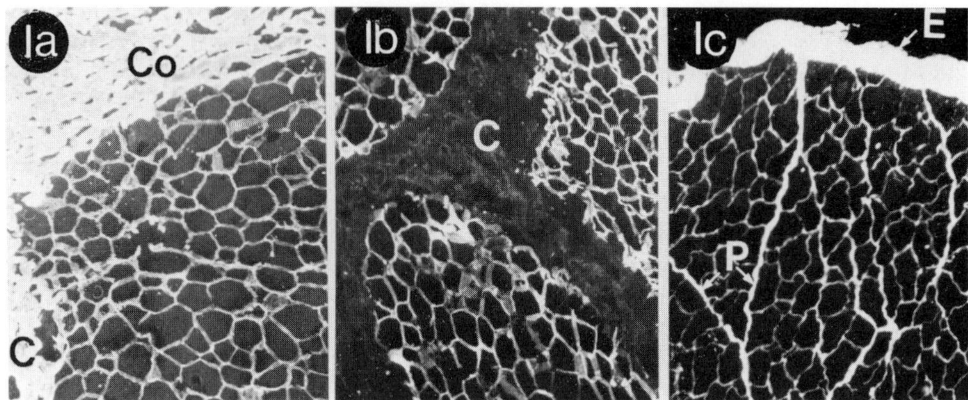


FIGURE 1. Indirect immunofluorescence of human s.c. adipose tissue (a,b) and adult chicken thigh muscle (c); frozen sections are stained with guinea pig antibodies to chicken Type I collagen (a,c) or to mouse Type IV collagen (b) and the appropriate fluorescent-conjugated anti-immunoglobulin. (From von der Mark, K., *Int. Rev. Connect. Tissue Res.*, 9, 298, 1981. With permission from Academic Press, N.Y.)

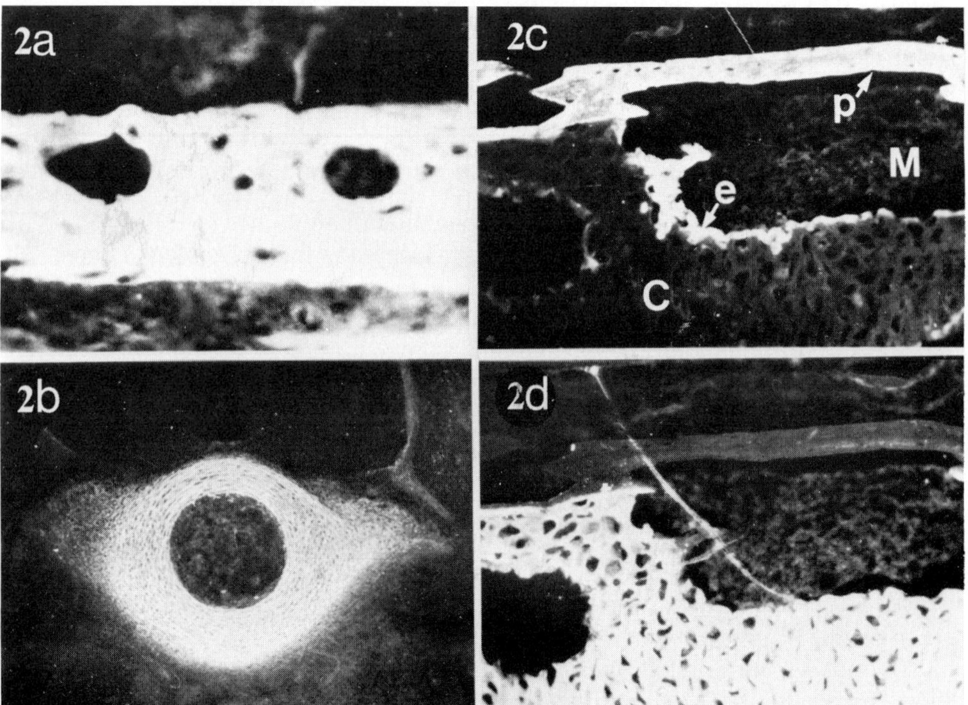


FIGURE 2. Indirect immunofluorescence of embryonic chicken long bone (17 days), decalcified and stained with rabbit antibodies to chicken Type I collagen (a). The vertebral cartilage of 7-day-old chicken embryos in (b) was pretreated with testicular hyaluronidase and stained with antibodies to Type II collagen. In 2c and 2d, sections from chicken tibia of 15-day-old embryos are shown which are stained with guinea pig antibodies of chicken Type I (c) and rabbit antibodies to chicken Type II (d) collagen. C, cartilage; M, marrow; p, periosteal bone; e, endochondral bone. (From von der Mark, K., *Int. Rev. Connect. Tissue Res.*, 9, 298, 1981. With permission from Academic Press, N.Y.)

articular surface by immunofluorescence[34] as well as biochemically.[102] In the growth plates of long bones, spicules of calcified cartilage are covered with layers of endochondral bone staining for Type I collagen.[31-34]

In subperichondral regions, in areas of endochondral ossification as well as in areas of tendon and ligament insertions into epiphyseal cartilage, frequently Types I and II collagen codistribute, as demonstrated by immunofluorescence double staining.[34]

A mixture of Types I and III collagen is also present in the intervertebral disk (see Figure 2b). A continuous gradient from Type I to Type II collagen has been demonstrated by immunofluorescence[33,35,37] as well as biochemically. The anulus fibrosis in addition contains Type III collagen.[35]

Pathological alterations of articular cartilage may be associated with the appearance of Types I and III collagen in the cartilage matrix. Conflicting data exist on the collagen types found in osteoarthritic cartilage. Deshmukh and Nimni[103] have reported that pieces of osteoarthrotic cartilage in culture synthesize Type I collagen. On the other hand, a study by Fukae et al.[104] suggested that the only collagen type present in osteoarthritic cartilage is Type II collagen; yet, Types I and III collagen were observed in the cartilage matrix of osteoarthrotic joints by immunofluorescence[31] and with biochemical methods. Different degrees of progression of the disease and sampling problems could account for these contrary reports.

### 6. *Tooth*

The presence of Type I collagen and the absence of Type III collagen in predentin and mineralized dentin of the unerupted rat incisor could be demonstrated by immunofluorescence.[64,65] Type III procollagen is present in embryonic dental mesenchyme, but disappears during odontoblast differentiation and reappears only with advancing vascularization of the dental papilla. Type IV is located in the capillaries and between inner and outer epithelium.[64,105] These data are well supported by biochemical studies and organ culture experiments (see Reference 106).

### 7. *Skeletal Muscle*

Immunohistological studies by Duance et al.[147] suggested that the outer connective tissue capsule of skeletal muscle, the epimysium, consists predominantly of Type I collagen, while the connective tissue fibers wrapping bundles of muscle fibers (perimysium), stained for Type III, but not for Type I collagen. The endomysial collagen investing individual muscle fibers stain for Types V and IV collagen.[30,51,52]

Other data, however, suggest that endomysium and perimysium also contain Type I collagen (see Figure 1c). The appearance and distribution of collagen Types I and III during development of skeletal muscle in the embryonic chick wing is described in an immunohistological study by Shellswell et al.[54]

### 8. *Heart*

Types I, III, and V collagen are found in the endomysium and perimysium, and Types IV and V are found in association with capillaries (see Figure 3) and possibly in the space around the cells. In embryonic heart muscle, the "major" collagen is Type I, but Types III and V were also identified by immunohistological means.

### 9. *Blood Vessels*

Extraction of collagen and studies on the biosynthesis of cells from various vascular beds suggest that the wall of larger blood vessels contains all the well-known collagen types, with the exception of Type II.[106] In addition, a high molecular weight

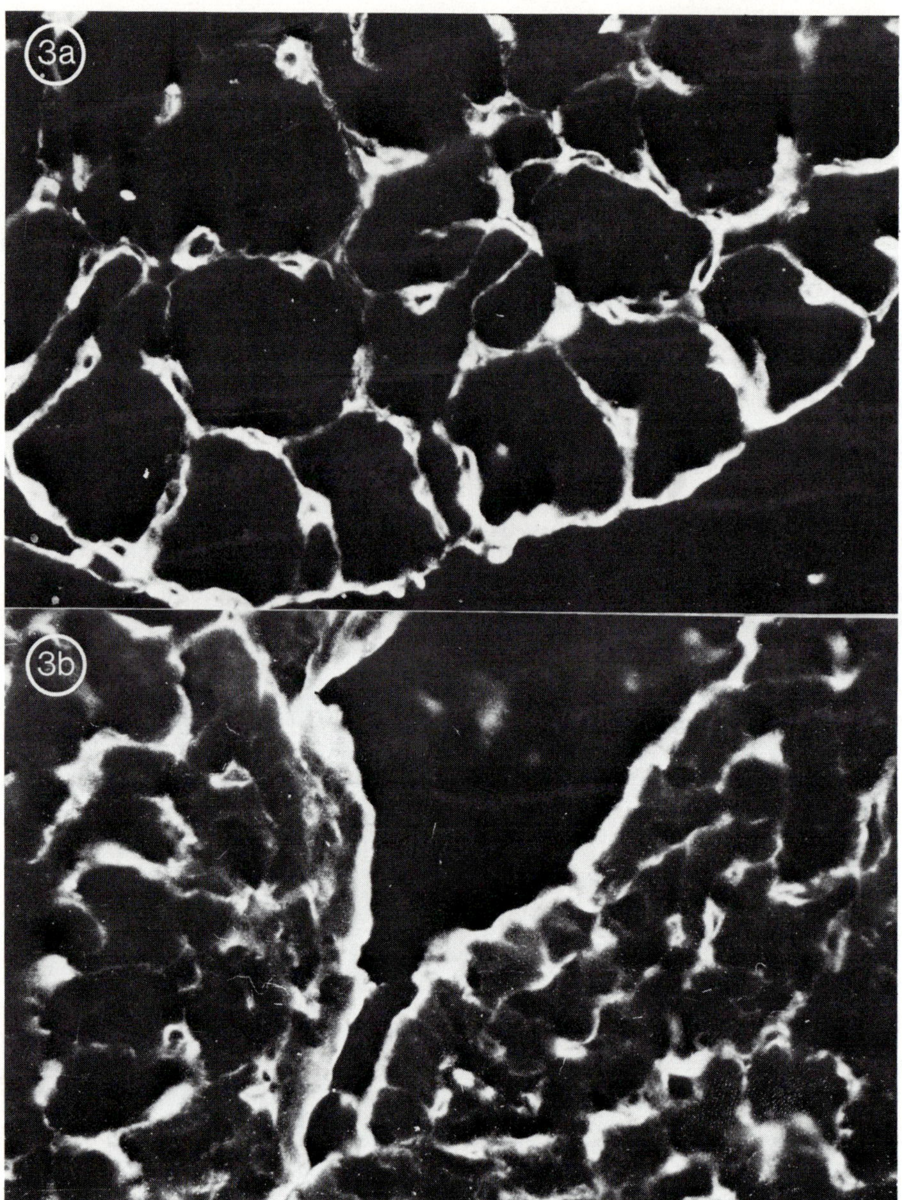


FIGURE 3. Indirect immunofluorescence of mouse heart muscle. Frozen sections are stained with rabbit antibodies to Type V (a), Type IV (b), and Type III (c) collagen. Type IV and V collagens are found in the wall of capillaries (notice the small circles in a and b) and in the subendocard (b), while Type III is only associated with larger fibrillar bands of matrix (c). (Courtesy of J. A. Madri, Yale University)

collagenous fragment has been isolated which resembles basement membrane collagens by amino acid composition,[107] and the vessel wall may contain the so-called EC-collagen synthesized by endothelial cells.[108]

By immunofluorescence collagen Types I, III, IV, and V are found throughout the media of the aortic wall.[8] The intima contains, however, only Types IV and V, and judged by this criterion, it resembles thus other vascular basement mem-

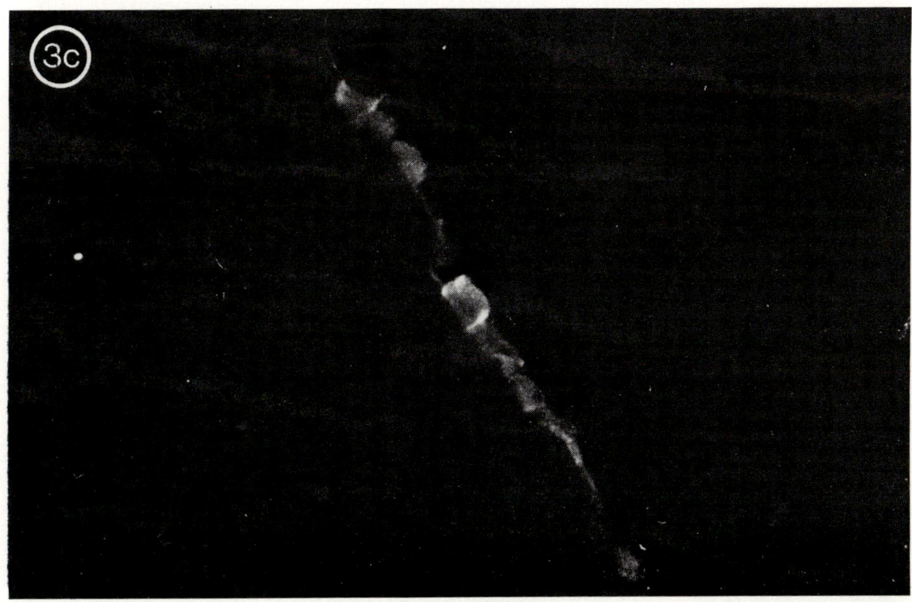


FIGURE 3c

branes.[8,43] As mentioned in the previous section, these data are in contrast to an earlier study[6] which describes Type III collagen at this site. Type III collagen in general is not seen in association with basement membranes (see Figure 4).

No ultrastructural data are available to confirm the structural arrangement of the collagens in more detail. By light microscopy, it appears that Types IV and V are within the basal laminae which surround individual smooth muscle cells.[8,39] This is consistent with the synthetic capability of smooth muscle cells. In vitro, smooth muscle cells produce, however, also other collagens.[8]

*10. Liver*

The results on content, distribution, and types of collagen present in the liver are confusing at present. Type III has been found immunohistochemically by several authors in liver vessels and intralobularly in the perisinusoidal space and in the periportal field.[11,45–49] In some studies, antibodies to Type I failed to detect this collagen.[11] More recent work detected it in the periportal field,[49] but not in the sinusoids. Other studies, however, clearly show a reaction of Type I antibodies along sinusoidal walls.[45,106] Type IV collagen is located within the liver lobules in the sinusoids, in blood and lymph vessels, in bile ducts and ductules and around nerves,[48] and codistributes with Type V.[143]

Attempts have been made to quantitatively describe by immunofluorescence the changes in collagen as a result of pathobiological events taking place in the liver.[11,47,53] In view of the large increase in connective tissue in fibrosis or cirrhosis of the liver and the biochemical evidence supporting this increase in collagen content, there is no doubt that changes have occurred and that some of these changes are reflected in fluorescence intensity. It remains, however, questionable whether immunofluorescence results can be applied to evaluate such changes quantitatively.

Some of the discrepancies mentioned above apparently are due to species-specific differences. Rodent liver contains less collagen to start with in comparison to human. Roll and Furthmayr[143] have shown that Types I and III, while present within the

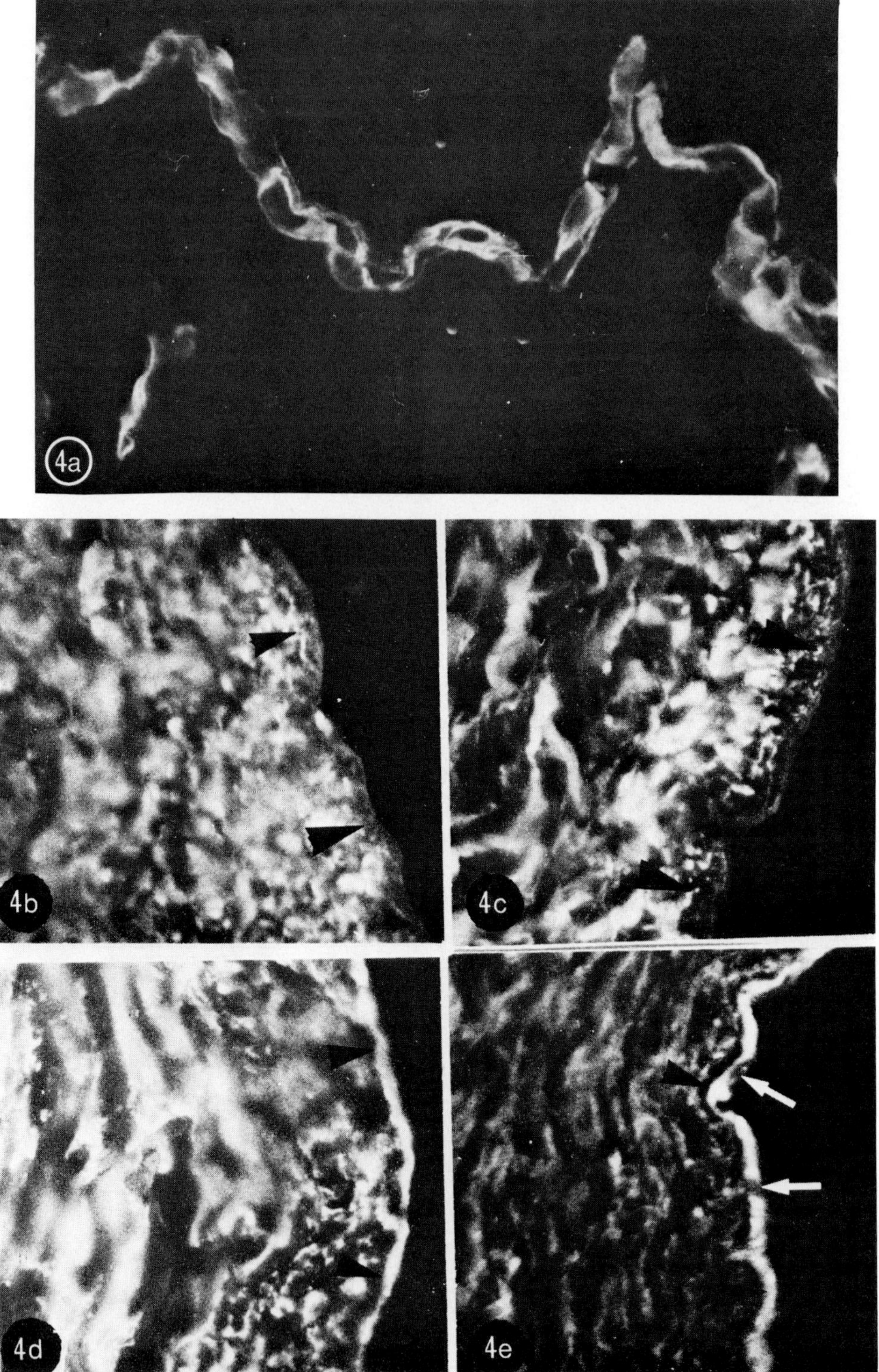


FIGURE 4. Indirect immunofluorescence of mouse lung (a) and aorta (b to e). Frozen sections were stained with rabbit antibodies of Type IV (a,d), Type I (b), Type III (c), or Type V (e) collagen. The alveolar and capillary basement membranes of the lung contain types IV and V (not shown) collagen in indistinguishable staining patterns. The wall of the aorta essentially contains all the collagens, but notice the absence or low degree of staining in the subendothelium with antibodies to Types I and III collagen (b,c) and staining in this area with Types IV and V antibodies (d,e). (Courtesy of J. A. Madri, Yale University)

FIGURE 5. Indirect immunofluorescence of mouse liver. Frozen sections of liver are stained with antibodies to Type I (a), Type III (b), Type IV (c), and Type V (d) collagen. The sinusoids in the liver apparently contain Type IV and possibly Type V collagen. Both of these collagens are also associated with the central artery (c, top left) and the structures of the portal triad (d, center right), while Types I and III within the lobules of the liver appear to stain only sparsely short fibrillar material.

sinusoids in rat and mouse liver, are found only in the form of short fluorescent strands here and there (see Figure 5). The fine reticular meshwork which becomes apparent with silver stains in rodents is much more delicate than in human liver. It could correspond to Types IV and V and possibly other glycoproteins which underly the sinusoidal endothelial cells to give support to the cells. More systematic work will be required to understand the contribution of liver cells and mesenchymal cells, particularly for the changes occurring during fibrotic reactions.

*11. Lung*

The studies on connective tissue of the lung more or less mirror the accumulation of knowledge in other organ systems. Due to the extreme insolubility of tissue collagen, biochemical studies initially did not recognize the complexity,[109] but with the discovery of the various collagen types in other tissues, eventually all of them

FIGURE 5b

were found also in the lung.[39,41] As in other organs, Types IV and V are associated with capillary and alveolar basement membranes, (see Figure 4A). Types I and III are found within septa and apparently scattered within segments of the alveolar walls, and Type II is seen in the cartilage of the tracheobronchial tree.

The biochemical data on collagen content, distribution of types, and even the apparent increase of collagen in fibrosis[76] of the lung are quite variable, and it is questionable whether studies using biochemical extraction techniques do justice to the complexity of lung architecture and can provide meaningful data.

*12. Kidney*

Because of the important physiological role of the glomerular basement membrane as a barrier for filtration and as a site for pathological reactions, much work is being devoted to understand the structural organization of this particular basement membrane. Glomeruli from steer kidney had been one of the first sources for basement membrane collagen.[110] However, this area of research has suffered from difficulties of obtaining enough material and of solubilizing basement membranes. Much of the information accumulated on basement membrane collagens isolated from other tissue sources will obviously be of value to eventually provide the necessary background for approaching the difficult problem of basement membrane structure and organi-

FIGURE 5c

zation in the glomerules.[111] It has been shown, for instance, that antibodies to Type IV collagens isolated from a mouse tumor, human placenta, or other sources stain also the basement membranes of renal glomeruli and tubuli.[39,43] This could indicate that the same molecules are used as building blocks for a variety of basement membranes. However, considerably more information on the chemical nature and possible heterogeneity of basement membrane collagens will be required before valid conclusions can be reached. Glomerular and tubular basement membranes also contain Type V collagen by immunofluorescence light microscopy (see Figure 6) and by immunoelectron microscopy[43] in addition to other glycoproteins.[42,44,112] The interstitial collagen Types I and III are found in the connective tissue matrix outside glomerular structures.[3,95]

*13. Placenta*

This organ is at present the main source for the preparation of Types IV and V collagen, but it also contains considerable amounts of Types I and III collagen. Type IV collagen is preferentially extracted from the chorionic plate after it has been stripped free of the placental membrane, but also from whole placenta. Only A and B chains, or $\alpha 2(V)$, $\alpha 1(V)$, of Type V collagen are found associated with the amnionic and chorionic membranes, while extracts of whole placenta also contain the C chain,

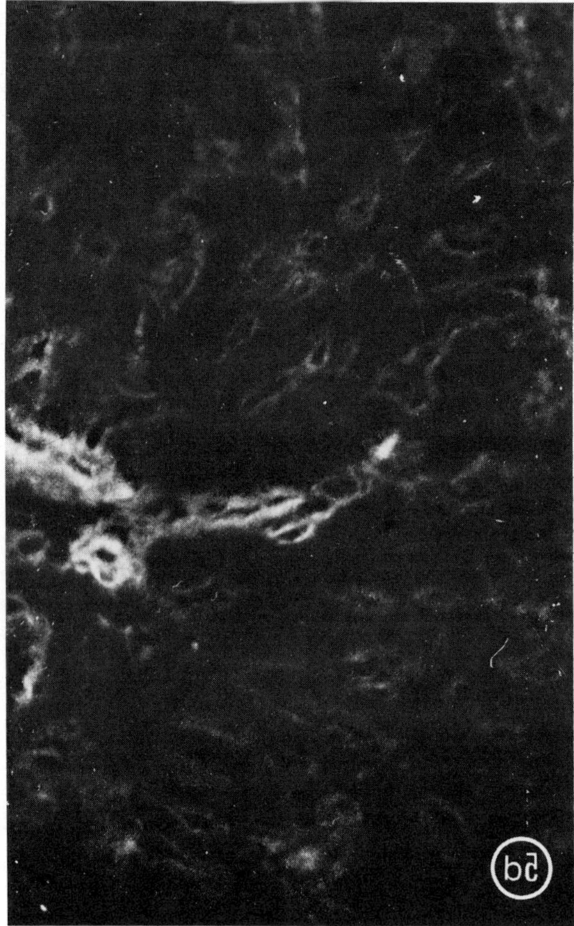

FIGURE 5d

or α3(V), in addition to A and B chains.[113] No systematic study of the architecture of placenta has been made with immunohistological techniques except for one study on placental membranes[67] which concluded that the types of collagen synthesized by amnionic cells in vitro are detectable in vivo in the various layers. Since placenta is rich in vascular tissue, it comes as no surprise that the basement membranes contain Types IV and V.[132]

*14. Eye*

The vertebrate eye is one of the most interesting and rewarding subjects for studies on the organization and function of connective tissue since within a rather small compartment of the organism several distinct tissues are organized in close proximity, such as the lens capsule, vitreous humour, the cornea, or neural retina. For this and other reasons, not the least being the interest in studying the development of this organ, much information is available. Initial biochemical evidence for a possible heterogeneity of collagen types in the avian eye was obtained by Trelstad and Kang,[114] who analyzed cornea, lens, vitreous body, and sclera from 1500 lathyritic chicken. They described Type I collagen in cornea and fibrous sclera, while the lens capsule and vitreous body suggested the presence of a different collagen consisting of only one α-chain.

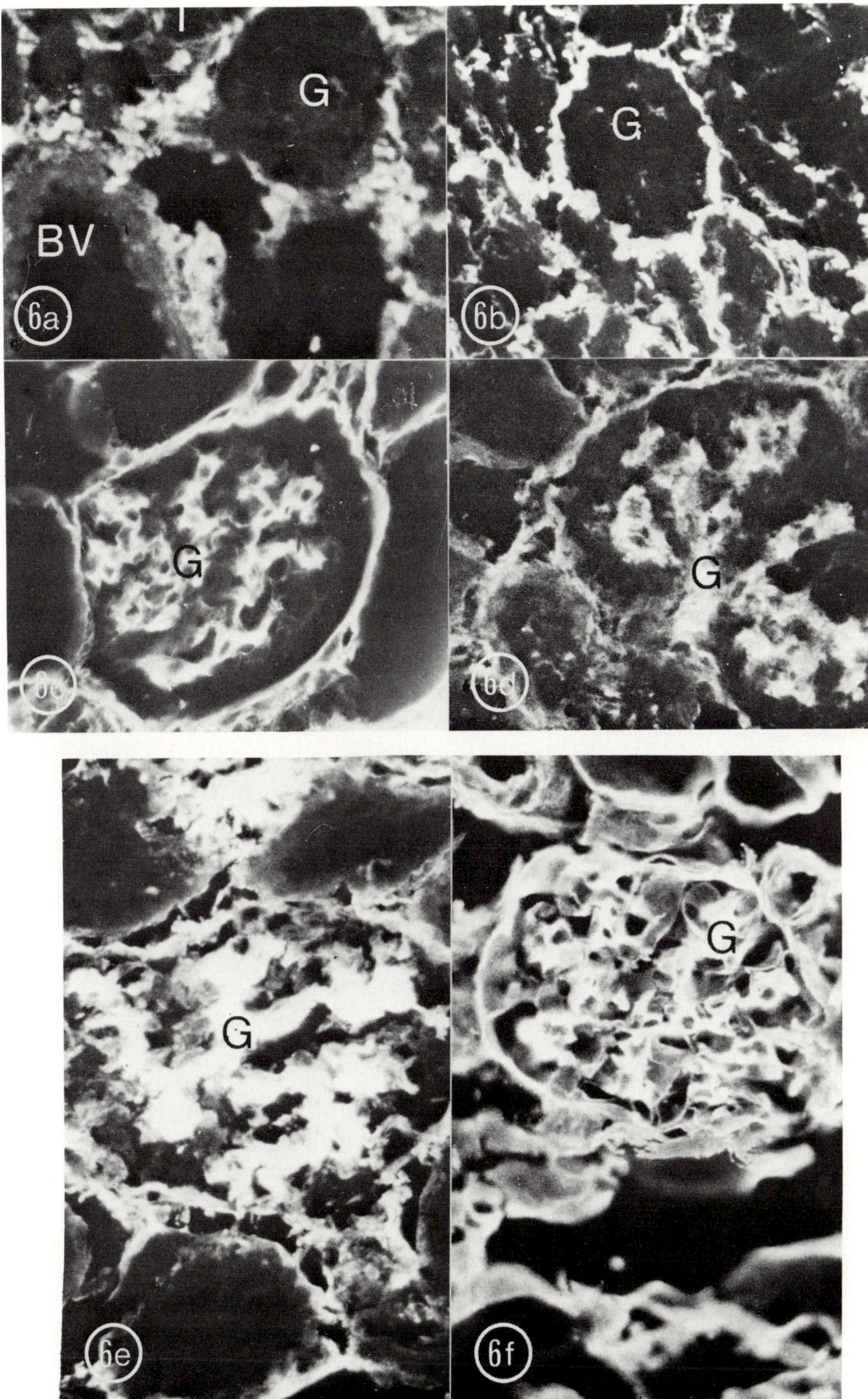


FIGURE 6. Indirect immunofluorescence of the mouse kidney. Antibodies to Type I (a), Type III (b), Type IV (c), and Type V (d) collagen and fibronectin (e) and laminin (f) are used to stain frozen sections. G, glomerulus; T, tubule; BV, blood vessel. Notice the absence of staining for Types I and III within the glomerulus. It is difficult to resolve unambiguously staining patterns (basement membrane vs. mesangium) within the glomerulus in the case of fibronectin, laminin, Type IV, or Type V with the light microscope, although these are consistent for the latter three with localization in basement membranes.

Type I collagen was found later to constitute the major collagen type in cornea not only of the chicken, but also of cattle, rabbit, or shark (reviewed in References 59 and 106). Immunofluorescence studies by von der Mark et al.[58] demonstrated that Type II collagen appears in the primary stroma at Stage 20 (2½ days of gestation) and persists until Stage 35 (14 days) in the anterior part of the secondary stroma, as well as in Descemet's membrane.[59] It is not found in the cornea of adult chicken. Antibodies to Type IV collagen stain Descemet's membrane and Bowman's membrane in the cornea of chick, mouse, human, and calf, and this also appears to be the only collagen present in the lens capsule.[144]

Biochemical and immunohistological studies on the collagen in chick vitreous humor suggested that it is related to Type II collagen[59,88,114,118] although it shows an electrophoretic mobility slightly different from α1(II) obtained from articular cartilage.[118]

Type II collagen in the vitreous humor in earlier embryonic stages apparently originates from the neural retina which secretes not only Type II collagen, but also Type V collagen into the vitreous body.[59,88,89,119] In later stages of development, Type II collagen seems to be also produced by hyalocytes.[88] In the avian eye, Type II collagen is synthesized by the pigmented epithelium, and it is present in scleral cartilage.[59]

*15. Nerves*

In peripheral human femoral nerve, Types I and III collagen were found in a ratio of 81 : 19.[121] Shellswell et al.[54] identified in addition Types IV and V collagen in the endoneurium and perineurium of bovine sciatic nerve by SDS electrophoresis and immunofluorescence.

*16. Collagen Types in Early Embryonic Development*

During embryonic development of many organs and tissues, a rapid transition of collagen types seems to occur which follows a highly complex schedule. In early stages of development, collagen types are frequently replaced within hours by other collagens, suggesting not only precise control mechanisms of collagen type synthesis and degradation, but also distinct functions for each collagen type.

The first collagen type synthesized in embryonic development appears to be Type IV collagen; it is localized by immunofluorescence in the inner cell mass of preimplantation mouse blastocysts.[7,9]

In the postimplantation mouse embryo at the fifth day of gestation, Type IV collagen was localized with the immunoperoxidase method in the primitive endoderm and subsequently in the parietal endoderm, Reichert's membrane, and between visceral endoderm and ectoderm. The synthesis of Type IV collagen by parietal endoderm of mouse[56] and rat[122] has been confirmed with biochemical techniques.

Type I collagen appears first in the four somite mouse embryo in mesodermal tissues, e.g., in head mesenchyme and heart mesenchyme and preferentially along basement membranes.[7] Similar patterns of distribution have been observed in the early chick embryo.[12]

A report[9] in which Type III collagen and the B chain of Type V collagen was found by immunofluorescence in the inner cell mass of preimplantation blastocysts could not be confirmed by Leivo et al.,[7] who presented evidence that Type III (pro)collagen appears in the mouse embryo at the same developmental stages and the same sites as Type I collagen.

Type II collagen is first seen in the notochord sheath of a 2-day-old (Stage 15) chick embryo[12] and subsequently in the primary corneal stroma at the basal surface of the neural retina[58,89] and in the vitreous humor.[88] Cartilaginous Type II collagen finally appears in the limb bud cartilage blastema and in the vertebral body;[12,34,57] for review see Reference 123.

## IV. LOCALIZATION OF CONNECTIVE TISSUE GLYCOPROTEINS IN DIFFERENT ORGANS

In addition to collagens, the extracellular matrix at different sites contains a number of high molecular weight glycoproteins. Although the existence of such proteins has been known for a long time,[124] it was not until fairly recently that distinct glycoproteins could be isolated and better defined. Three of these proteins have been fairly well characterized: fibronectin, laminin, and chondronectin. Short summaries of their chemical and biological properties can be found in Chapters 8–10 in Volume 1, and several extensive reviews have been published.[128,129] Antibodies to these proteins have been used extensively and have complemented the biochemical studies. They were applied to localize these connective tissue proteins in different organs in vivo to analyze cells which synthesize one or the other of these noncollagenous proteins and to seek the in vivo correlate for their biological function which has been mainly studied in tissue culture systems.

Fibronectin is synthesized in vitro mainly by fibroblasts and a number of other mesenchymal cells, such as smooth muscle cells, chondroblast, and myoblast precursor cells, but it is also made by endothelial cells, macrophages, and some epithelial cells. Laminin is made by a variety of epithelial cells, neuroblastoma cells, and endothelial cells. Chondronectin synthesis thus far has only been studied in cartilage cells. These glycoproteins are also present in plasma. Fibronectin is the most abundant of the three, but laminin and chondronectin are detectable with sensitive radioimmunoassays or biological assays.

Several questions come to mind when the localization of these noncollagenous proteins is studied in tissues, and some of the problems are not always realized:

1. Are these proteins deposited at the site of their synthesis or are they produced at a different site and translocated to sites at which they become detectable by immunohistological tests?
2. Are the protein antigens found at a certain site incorporated into the matrix and do they serve a critical function or do they merely represent "trapped" serum components?
3. Can one deduce from their localization which cells participate in their production?
4. What role do they play for the structural organization of connective tissue, e.g., the basement membrane or fibrillar interstitium?
5. Do in vitro studies on the synthesis of these proteins reflect their expression in vivo?
6. Is the in vivo expression related to developmental events?
7. Is the major biological role related to organization of cells into tissues, e.g., to mediate cell adhesion or to maintain a differentiated phenotype of a specialized cell? Or does the expression of these proteins provide only a marker for differentiation? What is the relationship between extracellular matrix and cell behavior?

### A. Fibronectin

The group of fibronectin proteins are closely related to proteins which can be distinguished as two major forms, a plasmaprotein (300 μg/mℓ) and a cell surface associated form. Although subtle differences in molecular weight and some other properties have been described for cell surface fibronectin from different cells and for plasma fibronectin,[115,125] little structural information is available. Varying degrees of glycosylation could account for the differences in molecular weight, and some evidence is forthcoming.[115] Plasma fibronectin, although different in its chemical properties from the cell surface associated form, nevertheless can be incorporated into the extracellular matrix synthesized by cells at least in tissue culture.[131] Fibronectin isolated from plasma of one species and added to cells in culture will become detectable in the cell layer matrix deposited by cells even from another species. The dual presence of the protein in plasma and on cell surfaces (also referred to as exogenous and endogenous fibronectin) poses obvious problems, when attempts are made to localize these proteins in tissue sections. Most currently used antisera do not distinguish between the exogenous and endogenous form, nor are they species-specific. In addition, trapping of serum fibronectin in tissue cannot be easily controlled for since the protein adheres readily to macromolecules, such as collagen, heparin, DNA, and possibly other structural elements. To control for adequate washes or perfusion of organs and to remove serum proteins as preparation for tissue sectioning and staining with immunohistochemical methods, other serum proteins such as albumin sometimes are used as markers. In nonperfused kidneys, the glomerular basement membranes stain readily with antibodies to albumin or fibronectin in a pattern consistent with its localization on or in glomerular basement membranes.[145] After perfusion, albumin staining disappears, but a low degree of labeling of the basement membrane with antibodies to fibronectin is retained.[44] Is the seemingly small amount of fibronectin along the basement membrane and underneath the foot processes of the glomerular epithelial cell and contact sites of the endothelial cells indicative for *in situ* synthesis and deposition of the protein? Can these data be taken as confirmation for a role of fibronectin in cell-matrix interaction, or is the protein remaining after perfusion (adequate to remove detectable albumin) merely trapped due to an affinity somewhat higher than albumin? These are difficult questions, and to answer them will require more careful work. It is also recognized that antisera to fibronectin (can) contain contaminating antibodies to unrelated proteins. These considerations may explain some of the discrepancies in the immunohistochemical findings on various tissues up to this date. Although it is recognized that fibronectin is for the most part found in connective tissue proper in the form of fibrillar arrays that may be distinct from collagenous fibrils or that may contain certain types of collagen, the association of fibronectin with basement membranes is more controversial. Fibronectin appears to be readily detectable in association with vascular basement membranes in some studies on chicken or mouse embryos[132,133] and in human organs.[134] Data have also been presented demonstrating labeling of subepithelial basement membranes of the small intestine by immunofluorescence[134,135] or of basement membranes in glandular structures. Other vascular beds such as the glomerular capillaries (See Figure 6E) in the kidney[112,134] or other basement membranes such as in skin apparently do not contain fibronectin, although results contradicting this statement can be found in the literature.[44] Most data, although almost exclusively obtained by immunofluorescence light microscopy, have been readily interpreted to support the notion of a general role of fibronectin in cell adhesion to matrix components. Some of the caveats have been mentioned already. Rarely have adequately controlled ultrastructural studies been done which would allow to distinguish between localization within, on, or at the cell-matrix interphase. In the case of glomerular basement membranes, ultrastructural studies have not been able as yet to resolve the issue.

The controversial finding of both presence and absence of fibronectin in association with, e.g., the vascular basement membrane, could be interpreted in favor of either differences in endothelial cells at the various sites or in favor of a role of fibronectin of lesser importance than suggested from the in vitro work.

### B. Laminin

Much less information is available for laminin and its localization in organs. It is found associated with all basement membranes studied thus far — skin, glomerular (See Figure 6F) and tubular basement membranes in the kidney, heart, and vascular tissue.[126,129] Ultrastructural studies on the localization of this large glycoprotein seem to indicate that it is present within the basement membrane proper of the renal tubular,[112] but the lamina lucida of the renal glomerular and skin basement membrane.[112,136]

## V. CONCLUSIONS

Since the discovery of the genetic heterogeneity of collagen in the pioneering work of Kefalides,[110] Miller and Matukas,[84] and Trelstad et al.,[101] numerous studies have dealt with the localization of known collagen types in tissues and the search for new collagen types for various reasons. From the beginning, one of the intriguing question was concerned with the function of the different collagen types. The discrete location of collagen types in different tissues, the rapid turnover of collagen types and replacement by other collagen types in embryonic development, and recent in vitro observations on preferential attachment of a number of cell types to distinct collagen types[137,138] strongly indicate a biological role of the individual collagens, beyond that expected of a merely structural molecule. Besides their function as supporting element of extracellular matrices, collagens seem to be involved in a number of cellular events; they apparently serve as substrates for cell adhesion, cell migration, and spreading, and they could influence cell proliferation as well as cell differentiation.

In general, collagen types seemed to be equally active in interactions with cells in vitro. However, recently more evidence became available which indicates that individual collagen types interact specifically with certain cells:

1. Collagen induced platelet aggregation is stimulated better by Type III collagen than with other collagen types[139]
2. Chondrogenic differentiation of somites is stimulated by Type II collagen better than with Types I or III collagen[140]
3. Certain cells attach with a preference to distinct collagen types, e.g., epidermal cells[137] and mammary epithelial cells to Type IV collagen,[138] chondrocytes to Type II collagen,[141] and smooth muscle cells to Type V collagen[146]

For the identification, localization, and quantative determination of collagen types, a number of sensitive and specific biochemical techniques have been developed. Immunohistological techniques have been used for localizing collagen types at the light microscopical and ultrastructural level, in particular in tissues or regions of tissues too small to allow biochemical analysis. Cell culture work has complemented many of the studies. It remains, however, for future efforts to refine and correlate this vast amount of information generated by the presently available technology to answer the many questions raised by these observations.

## ACKNOWLEDGMENTS

We wish to thank Dr. R. Timpl for a preprint of his article on "Antibodies to Collagens and Procollagens" and Dr. F. J. Roll and Dr. J. A. Madri for providing unpublished immunofluorescence data. The work in the laboratory of Heinz Furthmayr was supported by a contract from the National Cancer Institute #NO1-CB-84225 and grants from the Fairfield Chapter of the American Heart Association, the Anna Fuller Fund, NIH/DRR Basic Research Grants, and a Research Award from the American Cancer Society.

## REFERENCES

1. **Roll, F. J., and Madri, J. A.,** Immunocytochemical techniques in connective tissue research, in *Immunochemistry of the Extracellular Matrix,* Vol. 2, Furthmayr, H., Ed., CRC Press, Boca Raton, Fla., 1982, chap. 3.
2. **Watson, R. F., Rothbard, S., and Vanamee, P.,** The antigenicity of rat collagen, *J. Exp. Med.,* 99, 535, 1954.
3. **Wick, G., Furthmayr, H., and Timpl, R.,** Purified antibodies to collagen: an immunofluorescence study of their reaction with tissue collagen, *Int. Arch. Allergy Appl. Immunol.,* 48, 664, 1975.
4. **Timpl, R., Glanville, R. W., Wick, G., and Martin, G. R.,** Immunochemical study on basement membrane (type IV) collagens, *Immunology,* 38, 109, 1979.
5. **Furthmayr, H.,** Immunization procedures, isolation by affinity chromatography and serological and immunochemical characterization of collagen specific antibodies, in *Immunochemistry of the Extracellular Matrix,* Vol. 1, Furthmayr, H., Ed., CRC Press, Boca Raton, Fla., 1982, chap. 11.
6. **Gay, S., Balleisen, L., Remberger, K., Fietzek, P. P., Adelmann, B. C., and Kühn, K.,** Immunohistochemical evidence for the presence of collagen type III in human arterial walls, arterial thrombi and in leucocytes incubated with collagen in vitro, *Klin. Wochenschr.,* 53, 899, 1975.
7. **Leivo, I., Vaheri, A., Timpl, R., and Wartiovaara, J.,** Appearance and distribution of collagens and laminin in the early mouse embryo, *Dev. Biol.,* 76, 100, 1980.
8. **Madri, J. A., Dreyer, B., Pitlick, F. A., and Furthmayr, H.,** The collagenous components of the subendothelium, *Lab. Invest.,* 43, 303, 1980.
9. **Sherman, M. I., Gay, R., Gay, S., and Miller, E. J.,** Association of collagen with preimplantation and peri-implantation mouse embryos, *Dev. Biol.,* 74, 470, 1980.
10. **Foellmer, H., Madri, J. A., and Furthmayr, H.,** Monoclonal antibodies to type IV collagen from human placenta and calf lung, in preparation.
11. **Gay, S., Fietzek, P. P., Remberger, K., Eder, M., and Kühn, K.,** Liver cirrhosis: immunofluorescence and biochemical studies demonstrate two types of collagen, *Klin. Wochenschr.,* 53, 205, 1975.
12. **von der Mark, H., von der Mark, K., and Gay, S.,** Study of differential collagen synthesis during development of the chick embryo by immunofluorescence. I. Preparation of collagen type I and type II specific antibodies and their application to early stages of the chick embryo, *Dev. Biol.,* 48, 237, 1976.
13. **Vertel, B. M. and Dorfman, A.,** An immunochemical study of extracellular matrix formation during chondrogenesis, *Dev. Biol.,* 62, 1, 1978.
14. **Meigel, W. M., Gay, S., and Weber, L.,** Dermal architecture and collagen type distribution, *Arch. Dermatol. Res.,* 259, 1, 1977.
15. **Epstein, E. H., and Munderloh, N. H.,** Human skin collagen. Presence of type I and type III at all levels of the dermis, *J. Biol. Chem.,* 253, 1336, 1978.
16. **Fleischmajer, R., Perlish, J. S., Krieg, T., and Timpl, R.,** Variability in collagen and fibronectin synthesis by scleroderma fibroblasts in tissue culture, *J. Invest. Dermatol.,* 76, 400, 1981.
17. **Schauenstein, K. and Wick, G.,** *Acta Histochem. Suppl.,* 22, 101, 1980.
18. **Gay, S., Martin, G. R., Müller, P. K., Timpl, R., and Kühn, K.,** Simultaneous synthesis of types I and III collagen by fibroblasts in culture, *Proc. Natl. Acad. Sci. U.S.A.,* 73, 4073, 1976.
19. **Herrmann, H., Dessau, W., Fessler, L. I., and von der Mark, K.,** Synthesis of types I, III and $AB_2$ collagen by chick tendon fibroblasts in vitro, *Eur. J. Biochem.,* 105, 63, 1980.

20. **Novikoff, A. B., Novikoff, P. M., Quintana, N., and Davis, C.,** Diffusion artifacts in 3,3′-diaminobenzidine cytochemistry, *J. Histochem. Cytochem.,* 20, 745, 1972.
21. **Kraehenbuhl, J. P. and Jamieson, J. D.,** Localization of intracellular antigens by immunoelectron microscopy, *Int. Rev. Exp. Pathol.,* 13, 1, 1974.
22. **Sternberger, L. C.,** *Immunocytochemistry,* John Wiley & Sons, New York, 1979.
23. **Pöschl, A. and von der Mark, K.,** Synthesis of type V collagen by chick corneal fibroblasts in vivo and in vitro, *FEBS Lett.,* 115, 100, 1980.
24. **Wick, G., Kraft, D., Kokoschka, E. M., and Timpl, R.,** Diagnostic application of specific antiprocollagen sera. I. Analysis of skin biopsies, *J. Clin. Immunol.,* 6, 182, 1976.
25. **Meigel, W., Gay, S., and Weber, L.,** Dermal architecture and collagen type distribution, *Arch. Dermatol. Res.,* 259, 1, 1977.
26. **Timpl, R., Martin, G. R., Bruckner, P., Wick, G., and Wiedemann, H.,** Nature of the collagenous protein in a tumor basement membrane, *Eur. J. Biochem.,* 84, 43, 1978.
27. **Fleischmajer, R., Gay, S., Perlish, J. S., and Cesarini, J. P.,** Immunoelectron microscopy of type III collagen in normal and scleroderma skin, *J. Invest. Dermatol.,* 75, 189, 1980.
28. **Yaoita, H., Foidart, J. M., and Katz, S. I.,** Localization of the collagenous component in skin basement membrane, *J. Invest. Dermatol.,* 70, 191, 1978.
29. **Becker, U., Nowack, H., Gay, S., and Timpl, R.,** Production and specificity of antibodies against the aminoterminal region in type III collagen, *Immunology,* 31, 57, 1976.
30. **Duance, V. C., Restall, D. J., Beard, H., Bourne, F. J., and Bailey, A. J.,** The location of three collagen types in skeletal muscle, *FEBS Lett.,* 79, 248, 1977.
31. **Gay, S., Müller, P. K., Lemmen, C., Remberger, K., Matzen, K., and Kühn, K.,** Immunohistological study on collagen in cartilage-bone metamorphosis and degenerative osteoarthrosis, *Klin. Wochenschr.,* 54, 969, 1976.
32. **von der Mark, K. and von der Mark, H.,** The role of three genetically distinct collagen types in endochondral ossification and calcification of cartilage, *J. Bone Jt. Surg.,* 59b, 458, 1977.
33. **Remberger, K. and Gay, S.,** Immunochemical demonstration of different collagen types in the normal epiphyseal plate and benign and malignant tumors of bone and cartilage, *Z. Krebsforsch.,* 90, 95, 1977.
34. **von der Mark, K., von der Mark, H., and Gay, S.,** Study of differential collagen synthesis during development of the chick embryo by immunofluorescence. II. Localization of type I and type II collagen during long bone development, *Dev. Biol.,* 53, 153, 1976.
35. **Wick, G., Nowack, H., Hahn, E., Timpl, R., and Miller, E. J.,** Visualization of type I and type II collagens in tissue sections by immunohistologic techniques, *J. Immunol.,* 117, 298, 1976.
36. **Gay, S., Rhodes, R. K., Gay, R. E., and Miller, E. J.,** Collagen molecules comprised of $\alpha 1(V)$ chains ($\beta$-chains): an apparent localization in the exocytoskeleton, *Collagen Res.,* 1, 53, 1981.
37. **Beard, H. K., Ryvar, R., Brown, R., and Muir, H.,** Immunochemical localization of collagen types and proteoglycan in pig intervertebral discs, *Immunology,* 41, 491, 1980.
38. **Gay, S., Walter, P., and Kühn, K.,** Characterization and distribution of collagen types in arterial heterografts originating from the calf carotis, *Klin. Wochenschr.,* 54, 889, 1976.
39. **Wick, G., Glanville, R. W., and Timpl, R.,** Characterization of antibodies to basement membrane (type IV) collagen in immunohistological studies, *Immunobiology,* 156, 372, 1979.
40. **Timpl, R., von der Mark, K., and von der Mark, H.,** Immunochemistry and immunohistology of collagens, in *Biology of Collagen,* Viidik, A. and Vuust, J., Eds., Academic Press, New York, 1980, 211.
41. **Madri, J. A. and Furthmayr, H.,** Isolation and tissue localization of type $AB_2$ collagen from normal lung parenchyma, *Am. J. Pathol.,* 94, 323, 1979.
42. **Scheinman, J. I., Foidart, J. M., and Michael, A. F.,** The immunohistology of glomerular antigens. V. The collagenous antigen of the glomerulus, *Lab. Invest.,* 43, 373, 1980.
43. **Roll, F. J., Madri, J. A., Albert, J., and Furthmayr, H.,** Codistribution of collagen types IV and $AB_2$ in basement membranes and mesangium of the kidney, *J. Cell Biol.,* 85, 597, 1980.
44. **Courtoy, P. J., Kanwar, Y. S., Hynes, R. O., and Farquhar, M. G.,** Fibronectin localization in the rat glomerulus, *J. Cell Biol.,* 87, 691, 1980.
45. **Ott, U., Hahn, E., Moshudis, E., Bode, J. C., and Martini, G. A.,** Immunohistologischer Nachweis von Typ I-und Typ III-Kollagen in Leberbiopsien:frühe und späte Veränderungen bei alkoholischer Lebererkrankung, *Verh. Dtsch. Ges. Inn. Med.,* 83, 537, 1977.
46. **Nowack, H., Gay, S., Wick, G., Becker, U., and Timpl, R.,** Specific antibodies against type III collagen and procollagen, *J. Immunol. Methods,* 12, 117, 1976.
47. **Wick, G., Brunner, H., Penner, E., and Timpl, R.,** The diagnostic application of specific antiprocollagen sera. II. Analysis of liver biopsies, *Int. Arch. Allergy Appl. Immunol.,* 56, 316, 1978.
48. **Hahn, E., Wick, G., Pencev, D., and Timpl, R.,** Distribution of basement membrane proteins in normal and fibrotic human liver: collagen type IV, laminin and fibronectin, *Gut,* 21, 63, 1980.

49. **Biempica, L., Morecki, R., Wu, C. H., Giambrone, M.-A., and Rojkind, M.,** Immunocytochemical localization of type B collagen, *Am. J. Pathol.,* 98, 591, 1980.
50. **Grimaud, J. A., Druguet, M., Peyrol, S., Chevalier, O., Herbage, D., and el Badrawy, N.,** Collagen immunotyping in human liver. Light and electronmicroscopic study, *J. Histochem. Cytochem.,* 28, 1145, 1980.
51. **Bailey, A. J., Shellswell, G. B., and Duance, V. C.,** Identification and change of collagen types in differentiating myoblasts and developing chick muscle, *Nature,* 278, 67, 1979.
52. **Krieg, T., Timpl, R., Alitalo, K., Kurkinen, M., and Vaheri, A.,** Type III procollagen is the major collagenous component produced by a continuous Rhabdomyosarcoma cell line, *FEBS Lett.,* 104, 405, 1979.
53. **Gay, S. and Miller, E. J.,** *Collagen in the Physiology and Pathology of Connective Tissue,* G. Fischer, Stuttgart, 1978.
54. **Shellswell, G. B., Restall, D. J., Duance, V. C., and Bailey, A. J.,** Identification and differential distribution of collagen types in the central and peripheral nervous system, *FEBS Lett.,* 106, 305, 1979.
55. **Wick, G., Haller, M., Timpl, R., Cleve, H., and Ziegelmayer, G.,** Mummies from Peru. Demonstration of antigenic determinants of collagen in the skin, *Int. Arch. Allergy Appl. Immunol.,* 62, 76, 1980.
56. **Adamson, E. D. and Ayers, S. E.,** The Localization and synthesis of some collagen types in developing mouse embryos, *Cell,* 16, 953, 1979.
57. **Dessau, W., von der Mark, H., von der Mark, K., and Fischer, S.,** Changes in the patterns of collagens and fibronectin during limb-bud chondrogenesis, *J. Embryol. Exp. Morphol.,* 57, 51, 1980.
58. **von der Mark, K., von der Mark, H., Timpl, R., and Trelstad, R. L.,** Immunofluorescent localization of collagen types I, II, and III in the embryonic chick eye, *Dev. Biol.,* 59, 75, 1977.
59. **Hay, E. D., Linsenmayer, T. F., Trelstad, R. L., and von der Mark, K.,** Origin and distribution of collagens in the developing avian cornea, *Curr. Top. Eye Res.,* 1, 1, 1979.
60. **Barrach, H. J. and Angermann, K.,** Immunofluorescence as a tool in teratological research (localization of collagen types in fetal tissues by immunofluorescence), in *Methods in Prenatal Toxicology,* Neubert, D., Merker, H. J., and Kwasigroch, E., Eds., Thieme, Stuttgart, 1977, 332.
61. **Reddi, A. H., Gay, R., Gay, S., and Miller, E. J.,** Transitions in collagen types during matrix-induced cartilage, bone and bone marrow formation, *Proc. Natl. Acad. Sci. U.S.A.,* 74, 5589, 1977.
62. **Foidart, F. M. and Reddi, A. H.,** Immunofluorescent localization of Type IV collagen and laminin during endochondral bone differentiation and regulation by pituitary growth hormone, *Dev. Biol.,* 75, 130, 1980.
63. **Ekblom, P., Lehtonen, E., Saxen, L., and Timpl, R.,** Shift in collagen type as an early response to induction of the metanephric mesenchyme, *J. Cell Biol.,* 89, 276, 1981.
64. **Lesot, H., von der Mark, K., and Ruch, J. V.,** Immunofluorescent localization of collagen types I, III, IV in the embryonic mouse tooth germ, *C. R. Acad. Sci. Ser. D.,* 286, 765, 1978.
65. **Thesleff, I., Stenman, S., Vaheri, A., and Timpl, R.,** Changes in the matrix proteins fibronectin and collagen, during differentiation of mouse tooth germ, *Dev. Biol.,* 70, 116, 1979.
66. **Shellswell, G. B., Bailey, A. J., Duance, V. C., and Restall, D. J.,** Has collagen a role in muscle pattern formation in the developing chicken wing? I. An immunofluorescence study, *J. Embryol. Exp. Morphol.,* 60, 245, 1980.
67. **Alitalo, K., Kurkinen, M., Vaheri, A., Krieg, T., and Timpl, R.,** Extracellular matrix components synthesized by human amniotic epithelial cells in culture, *Cell,* 19, 1053, 1980.
68. **Fleischmajer, R., Gay, S., Meigel, W. N., and Perlish, J. S.,** Collagen in the cellular and fibrotic stages of scleroderma, *Arthritis Rheum.,* 21, 418, 1978.
69. **Fleischmajer, R., Dessau, W., Timpl, R., Krieg, T., Luderschmidt, C., and Wiestner, M.,** Immunofluorescence analysis of collagen, fibronectin, and basement membrane protein in scleroderma skin, *J. Invest. Dermatol.,* 75, 270, 1980.
70. **Wick, G., Olsen, B. R., and Timpl, R.,** Immunohistologic analysis of fetal and dermatosparactic calf and sheep skin with antisera to procollagen and collagen type I, *Lab. Invest.,* 39, 151, 1978.
71. **Gay, S., Kresina, T. T., Gay, R., Miller, E. J., and Montes, L. F.,** Immunohistochemical demonstration of basement membrane collagen in normal human skin and in psoriasis, *J. Cutan. Pathol.,* 6, 91, 1979.
72. **Wick, G., Hönigsmann, H., and Timpl, R.,** Immunofluorescence demonstration of type IV collagen and a noncollagenous glycoprotein in thickened vascular basal membranes in protoporphyria, *J. Invest. Dermatol.,* 73, 335, 1979.
73. **Thesleff, I., Barrach, H. J., Foidart, J. M., Vaheri, A., Pratt, R. M., and Martin, G. R.,** Changes in the distribution of type IV collagen, laminin, proteoglycan, and fibronectin during mouse tooth development, *Dev. Biol.,* 81, 182, 1981.

74. **Kent, G., Gay, S., Inouye, T., Bahu, R., Minick, O. T., and Popper, H.,** Vitamin A containing lipocytes and formation of type III collagen in liver injury, *Proc. Natl. Acad. Sci. U.S.A.*, 73, 3719, 1976.
75. **Wick, G., Timpl, R., and Crystal, R. G.,** in preparation.
76. **Madri, J. A. and Furthmayr, H.,** Collagen polymorphism in the lung: an immunochemical study of pulmonary fibrosis, *Hum. Pathol.*, 11, 353, 1980.
77. **Gay, S., Viljanto, J., Penttinen, R., and Raekallio, J.,** Collagen types in early phases of wound healing in children, *Acta Chir. Scand.*, 144, 205, 1978.
78. **Kurkinen, M., Vaheri, A., Roberts, P. J., and Stenman, S.,** Sequential appearance of fibronectin and collagen in experimental granulation tissue, *Lab. Invest.*, 43, 47, 1980.
79. **Gay, S. and Remberger, K.,** Rheumatoide arthritis: immunohistochemische Untersuchungen mit Antikörpern gegen Kollagen Typ I, II and III, *Verh. Dtsch. Ges. Pathol.*, 60, 290, 1976.
80. **Rhodes, R. K. and Miller, E. J.,** Physicochemical characterization and molecular organization of the collagen A and B chains, *Biochemistry*, 17, 3442, 1978.
81. **Jimenez, S. A., Yankowski, R., and Bashey, R. I.,** Identification of two new collagen α-chains in extracts of lathyritic chick embryo tendons, *Biochem. Biophys. Res. Commun.*, 81, 1298, 1978.
82. **Hong, B. S., Davison, P. F., and Cannon, D. J.,** Isolation and characterization of a distinct type of collagen from bovine fetal membranes and other tissues, *Biochemistry*, 18, 4278, 1979.
83. **Welsh, C., Gay, S., Rhodes, K., Pfister, R., and Miller, E. J.,** Collagen heterogeneity of normal rabbit cornea. I. Isolation and biochemical characterization of the genetically distinct collagens, *Biochem. Biophys. Acta*, 625, 78, 1980.
84. **Miller, E. J., and Matukas, V. J.,** Chick cartilage collagen: a new type of α1 chain not present in bone or skin of the species, *Proc. Natl. Acad. Sci. U.S.A.*, 64, 1264, 1969.
85. **Eyre, D. R. and Muir, H.,** The distribution of different molecular species of collagen in fibrous, elastic and hyaline cartilages of the pig, *Biochem. J.*, 151, 595, 1975.
86. **Linsenmayer, T. F., Trelstad, R. L., and Gross, J.,** The collagen of chick embryonic notochord, *Biochem. Biophys. Res. Commun.*, 53, 39, 1973.
87. **Miller, E. J. and Matthews, M. B.,** Characterization of notochord collagen as a cartilage-type collagen, *Biochem. Biophys. Res. Commun.*, 60, 424, 1974.
88. **Newsome, D. A., Linsenmayer, T. F., and Trelstad, R. L.,** Vitreous body collagen. Evidence for a dual origin from the neural retina and hyalocytes, *J. Cell Biol.*, 71, 59, 1976.
89. **Smith, G. N., Linsenmayer, T. F., and Newsome, D. A.,** Synthesis of type II collagen in vitro by embryonic chick neural retina tissue, *Proc. Natl. Acad.Sci. U.S.A.*, 73, 4420, 1976.
90. **Miller, E. J.,** Biochemical characteristics and biological significance of the genetically-distinct collagens, *Mol. Cell. Biochem.*, 13, 165, 1976.
91. **Linder, E., Vaheri, A., Ruoslahti, E., and Wartiovaara, J.,** Distribution of fibroblast surface antigen in the developing chick embryo, *J. Exp. Med.*, 142, 41, 1975.
92. **Epstein, E. H.,** $[\alpha 1(III)]_3$ Human skin collagen release by pepsin digestion and preponderance in fetal life, *J. Biol. Chem.*, 249, 3225, 1974.
93. **Clore, J. N., Cohen, I. K., and Diegelman, R. F.,** Quantitation of collagen types I and III during wound healing in rat skin, *Proc. Soc. Exp. Med.*, 161, 337, 1979.
94. **Shuttleworth, C. A., Forest, L., and Jackson, D.,** Comparison of the cyanogen bromide peptides of insoluble guinea pig skin and scar collagen, *Biochem. Biophys. Acta*, 379, 207, 1975.
95. **Timpl, R., Wick, G., Furthmayr, H., Lapière, C. M., and Kühn, K.,** Immunochemical properties of procollagen from dermatosparactic calves, *Eur. J. Biochem.*, 32, 584, 1973.
96. **Timpl, R., Wick, G., and Gay, S.,** Antibodies to distinct types of collagens and procollagens and their application to immunohistology, *J. Immunol. Methods*, 18, 165, 1977.
97. **Stenn, K. S., Madri, J. A., and Roll, F. J.,** Migrating epidermis produces $AB_2$ collagen and requires continual collagen synthesis for movement, *Nature*, 277, 229, 1979.
98. **Dehm, P. and Prockop, D. J.,** Time lag in the secretion of collagen by matrix-free tendon cells and inhibition of the secretory process by colchicine and vinblastine, *Biochim. Biophys. Acta*, 264, 375, 1972.
99. **Chung, E., Rhodes, R. K., and Miller, E. J.,** Isolation of three collagenous components of probable basement membrane origin from several tissues, *Biochem. Biophys. Res. Commun.*, 71, 1167, 1976.
100. **Barrach, H. J. and Angermann, K.,** in *Methods in Prenatal Toxicology*, Neubert, D., Merker, H. J., Kwasigroch, T. E., Eds., Thieme, Stuttgart, 1977, 332.
101. **Trelstad, R. L., Kang, A. H., Igarashi, S., and Gross, J.,** Isolation of two distinct collagens from chick cartilage, *Biochemistry*, 9, 4993, 1970.
102. **Seyer, J. M., Brickley, D. M., and Glimcher, M. J.,** The identification of two types of collagen in the articular cartilage of postnatal chickens, *Calcif. Tissue Res.*, 17, 43, 1974.
103. **Deshmukh, K. and Nimni, M.,** Isolation and characterization of cyanogen bromide peptides from the collagen of bovine articular cartilage, *Biochem. J.*, 133, 615, 1973.

104. **Fukae, M., Mechanic, G. L., Adamy, L., and Schwartz, E. R.,** Chromatographically different type II collagen from human normal and osteoarthritic cartilage, *Biochem. Biophys. Res. Commun.*, 67, 1571, 1975.
105. **Lesot, H., Osman, M., and Ruch, J. V.,** Immunofluorescent localization of collagens, fibronectin, and laminin during terminal differentiation of odontoblasts, *Dev. Biol.*, 82, 371, 1981.
106. **von der Mark, K.,** Localization of collagen types in tissues, *Int. Rev. Connect. Tissue Res.*, 9, 265, 1981.
107. **Furuto, D. K. and Miller, E. J.,** Isolation of a unique collagenous fraction from limited pepsin digests of human placental tissue. Characterization of one of the constituent polypeptide chains, *J. Biol. Chem.*, 255, 290, 1980.
108. **Sage, H., Pritzl, P., and Bornstein, P.,** A unique, pepsin-sensitive collagen synthesized by aortic endothelial cells in culture, *Biochemistry*, 19, 5747, 1980.
109. **Crystal, R.,** Lung collagen: definition, diversity and development, *Fed. Proc. Fed. Amer. Soc. Exp. Biol.*, 13, 2243, 1974.
110. **Kefalides, N. A.,** Isolation and characterization of the collagen from glomerular basement membranes, *Biochemistry*, 7, 3103, 1968.
111. **Timpl, R. and Martin, G. R.,** Components of basement membranes, in *Immunochemistry of the Extracellular Matrix*, Vol. 2, Furthmayr, H., Eds., CRC Press, Boca Raton, Fla., 1981.
112. **Madri, J. A., Roll, F. J., Furthmayr, H., and Foidart, J.-M.,** Ultrastructural localization of fibronectin and laminin in the basement membranes of the murine kidney, *J. Cell Biol.*, 86, 682, 1980.
113. **Bornstein, P. and Sage, H.,** Structurally distinct collagen types, *Annu. Rev. Biochem.*, 49, 957, 1980.
114. **Trelstad, R. L. and Kang, A.,** Collagen heterogeneity in the avian eye: lens, vitreous body, cornea and sclera, *Exp. Eye Res.*, 18, 395, 1974.
115. **Ruoslahti, E., Engvall, E., Hayman, E. G., and Spiro, R. G.,** Comparative studies on amniotic fluid and plasma fibronectin, *Biochem. J.*, 193, 295, 1981.
116. **Yue, B. J. Y. T., Baum, J. L., and Smith, B. D.,** Collagen-synthesis by cultures of stromal cells from normal human and keratoconus corneas, *Biochem. Biophys. Res. Commun.*, 86, 465, 1979.
117. **Harnisch, J. P., Buchen, R., Sinha, P. K., and Barrach, H. J.,** Ultrastructural identification of type I and II collagen in cornea of mouse by means of enzyme labeled antibodies, *Albrecht von Graefes Archiv. Klin. Exp. Ophthalmol.*, 208, 9, 1980.
118. **Swann, D. A. and Scotsman, J. S.,** The chemical composition of bovine vitreous-humor collagen fibers, *Biochem. J.* 185, 545, 1980.
119. **Linsenmayer, T. F. and Little, C. D.,** Embryonic neural retina collagen: in vitro synthesis of high molecular weight forms of type II plus a new genetic type, *Proc. Natl. Acad. Sci. U.S.A.*, 75, 3235, 1978.
120. **Newsome, D. A. and Kenyon, K. R.,** Collagen production in vitro by the retinal pigmented epithelium of the chick embryo, *Dev. Biol.*, 32, 387, 1973.
121. **Seyer, J. M. and Whitacker, J. N.,** Characterization of Type I and Type III collagens from human peripheral nerve, *Biochim. Biophys. Acta*, 492, 415, 1977.
122. **Minor, R. R., Strause, E. L., Koszalka, T. R., Brent, R. L., and Kefalides, N. A.,** Organ cultures of the embryonic rat parietal yolk sac. II. Synthesis, accumulation, and turnover of collagen and noncollagen basement membrane glycoproteins, *Dev. Biol.*, 48, 365, 1976.
123. **von der Mark, K.,** Immunological studies on collagen type transition in chondrogenesis, *Curr. Top. Dev. Biol.*, 14, 199, 1980.
124. **Anderson, J. C.,** Glycoproteins of the connective tissue matrix, *Int. Rev. Connect. Tissue Res.*, 7, 251, 1976.
125. **Yamada, K. M.,** Isolation of fibronectin from plasma and cells, in *Immunochemistry of the Extracellular Matrix*, Vol. 1. Furthmayr, H., Ed., CRC Press, Boca Raton, Fla., 1982, chap. 8.
126. **Foidart, J.-M., Timpl, R., Furthmayr, H., and Martin, G. R.,** Laminin, a glycoprotein from basement membranes, in *Immunochemistry of the Extracellular Matrix*, Vol. 1. Furthmayr, H., Ed., CRC Press, Boca Raton, Fla., 1982, chap. 9.
127. **Hewitt, A. T., Varner, H. H., and Martin, G. R.,** Isolation and properties of chondronectin, the chondrocyte attachment factor, in *Immunochemistry of the Extracellular Matrix*, Vol. 1. Furthmayr, H., Ed., CRC Press, Boca Raton, Fla., 1982, chap. 10.
128. **Yamada, K. M. and Olden, K.,** Fibronectins-adhesive glycoproteins of cell surface and blood, *Nature*, 275, 179, 1978.
129. **Timpl, R., Rohde, H., Risteli, L., Ott, V., Gehron-Robey, P., and Martin, G. R.,** Laminin, *Methods Enzymol.*, in press.

130. **Kleinman, H. K., Rohrbach, D. H., Terranova, V. P., Varner, H. H., Hewitt, A. T., Grotendorst, G. R., Wilkes, C. M., Martin, G. R., Seppa, H., and Schiffman, E.**, Collagenous matrices as determinants of cell function, in *Immunochemistry of the Extracellular Matrix*, Vol. 2, Furthmayr, H., Ed., CRC Press, Boca Raton, Fla., 1982, chap. 6.
131. **Hayman, E. G. and Ruoslahti, E.**, Distribution of fetal bovine serum fibronectin and endogenous rat cell fibronectin in extracellular matrix, *J. Cell Biol.*, 83, 255, 1979.
132. **Foellmer, H., Madri, J. A., and Furthmayr, H.**, Staining by immunofluorescence of basement membranes in various organs, unpublished.
133. **Wartiovaara, J., Stenman, S., and Vaheri, A.**, Changes in expression of fibroblast surface antigens (SFA) in induced cytodifferentiation and in heterokaryon formation, *Differentiation*, 5, 85, 1976.
134. **Stenman, S. and Vaheri, A.**, Distribution of a major connective tissue protein, fibronectin, in normal human tissues, *J. Exp. Med.*, 147, 1054, 1978.
135. **Anarono, A., Isselbacher, K. J., and Ruoslahti, E.**, Fibronectin synthesis by epithelial crypt cells of rat small intestine, *Proc. Natl. Acad. Sci. U.S.A.*, 75, 5548, 1978.
136. **Foidart, J. M., Bere, E. W., Yaar, M., Rennard, S. I., Gullino, M., Martin, G. R., and Katz, S. I.**, Distribution and immunoelectron microscopic localization of laminin, a non-collagenous basement membrane glycoprotein, *Lab. Invest.*, 42, 336, 1980.
137. **Murray, J. C., Stingl, G., Kleinman, H. K., Martin, G. R., and Katz, S. I.**, Epidermal cells adhere preferentially to type IV (Basement Membrane) collagen, *J. Cell Biol.*, 80, 197, 1979.
138. **Wicha, M. S., Liotta, L. A., Garbisa, S., and Kidwell, W. R.**, Basement membrane collagen requirements for attachment and growth of mammary epithelium, *Exp. Cell Res.*, 124, 181, 1979.
139. **Balleisen, L., Nowack, H., Gay, S., and Timpl, R.**, Inhibition of collagen-induced platelet aggregation by antibodies to distinct types of collagens, *Biochem. J.*, 184, 683, 1979.
140. **Lash, J. W. and Vasan, N. S.**, Somite chrondrogenesis in vitro. Stimulation by exogenous extracellular extracellular matrix components, *Dev. Biol.*, 66, 151, 1978.
141. **Hewitt, A. T., Kleinman, H. K., Pennypacker, J. P., and Martin, G. R.**, Identification of an adhesion factor for chondrocytes, *Proc. Natl. Acad. Sci. U.S.A.*, 77, 385, 1980.
142. **Herrmann, H.**, Unpublished.
143. **Roll, F. J.**, Unpublished.
144. **von der Mark, K. and Timpl, R.**, Unpublished.
145. **Courtney, P.**, Unpublished, 1981.
146. **Grotendorst, G. R., Seppa, H., Kleinman, H. K., and Martin, G. R.**, Attachment of smooth muscle cells to collagen and their migration to platelet derived growth factor, submitted.
147. **Duance, V. C., Restall, D. J., Beard, H., Bourne, F. J., and Bailey, A. J.**, Location of three collagen types in skeletal muscle, *FEBS Lett.*, 79, 248, 1977.

Chapter 5

# COMPONENTS OF BASEMENT MEMBRANES

**Rupert Timpl and George R. Martin**

## TABLE OF CONTENTS

# I. INTRODUCTION AND HISTORICAL BACKGROUND

In this article we will outline our current understanding of the components of basement membranes. Basement membranes are sheet-like, extracellular matrices found in many locations in the body which separate epithelial as well as endothelial cells from underlying stroma.[1] They were thought at one time to be thickened cell membranes. However, ultrastructural studies showed that they were outside the cell membrane and composed of an electron lucent zone adjacent to the cell, the lamina lucida, plus a denser zone known as the lamina densa or basal lamina[1] (see Figure 1).

Glomeruli and lens capsule have been frequently used as a source of basement membrane since these structures can be isolated free from other tissue elements.[2,3] Chemical analyses established that basement membranes contained collagenous protein as indicated by the presence of collagen-specific amino acids, such as hydroxyproline and hydroxylysine.[3,4] Since the glycine level was much lower than the 333 residues per 1000 amino acids present in interstitial collagens,[4,5] it appeared that the basement membranes contained a substantial proportion of noncollagenous protein. At the ultrastructural level, the basal lamina was observed as a felt-like network of fine fibrils[6] in contrast to other collagenous matrices where the fibers are larger, aligned, and exhibit periodicity.[7,8]

Due to the limited amounts of tissue isolated from such sources as glomeruli and lens capsule and the limited solubility of their constituents, it proved difficult to characterize basement membranes. However, during this period, it was discovered that distinct types of collagen were present in the body, including Type I collagen in bone, skin, tendon, etc., Type II collagen in cartilage, and Type III collagen in blood vessels and other sites.[8,9] By treating glomeruli or lens capsules with pepsin, Kefalides was able to solubilize collagenous protein, and he reported the isolation of a unique basement membrane collagen, Type IV, containing a single type of chain, the α1(IV) chain.[3] He suggested that basement membranes were composed of Type IV collagen linked to noncollagenous glycoproteins by disulfide bonds.[10]

Spiro and collaborators used denaturing solvents and reducing agents to extract collagenous proteins from isolated glomeruli.[5,11,12] The collagenous protein isolated in this fashion was found to be heterogeneous, with components ranging in size from 20,000 to over 200,000 daltons. Several such components were isolated and found to differ in composition from one another, most notably in the level of amino acids, such as glycine, hydroxyproline, and hydroxylysine, which are enriched in the helical portion of collagen. These investigators proposed that the glomerular basement membrane was composed of a number of dissimilar peptide subunits with collagenous and noncollagenous sequences of variable length.

Biosynthetic studies using basement membrane-producing tissues and cells have aided in establishing the nature of the collagenous protein. The initial studies showed a single chain estimated to be between 140,000 and 180,000 daltons[13,14] which associated to form the Type IV collagen molecule. These biosynthetic studies, however, underestimated the number of different chains, and more recent studies indicate that there are at least two different chains, α1(IV) and α2(IV), produced by basement membrane synthesizing cells (see Table 1). Subsequently, other collagenous components, such as Type V collagen,[15,16] 7-S collagen,[17] and intima collagen,[16,18] have been isolated from tissues with basement membranes and may also be components (see Table 1).

It was apparent from early histochemical studies that basement membranes con-

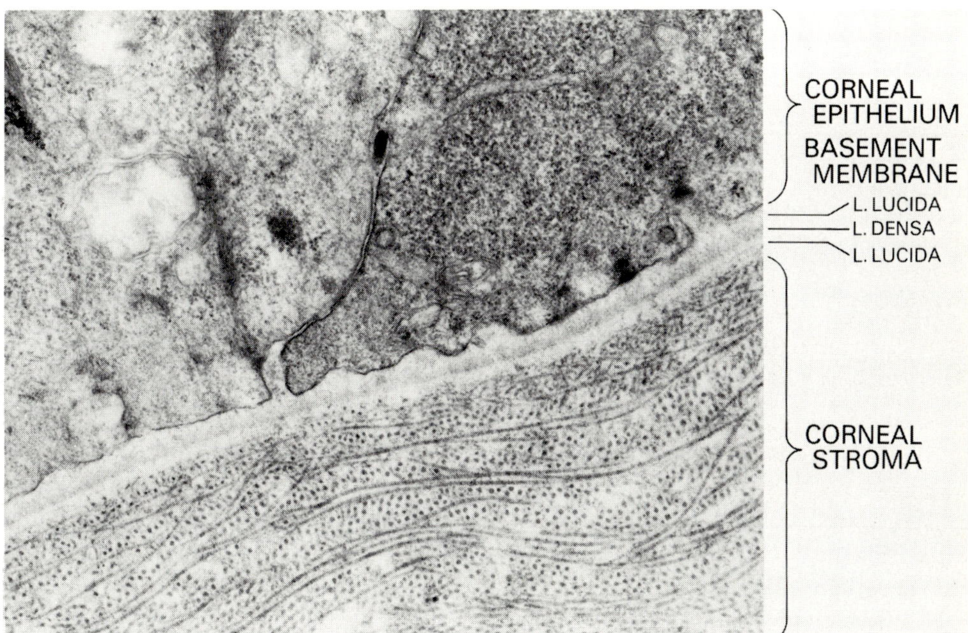


FIGURE 1. Micrograph of corneal epithelium and mesenchyme from developing chick eye. The zones of the basement membrane which separate mesenchyme from epithelium are indicated. (We are indebted to Dr. R. Trelstad for the picture.)

tained abundant carbohydrate[19] now thought to occur in glycoproteins. Two such glycoproteins, laminin and fibronectin, as well as a heparin sulfate proteoglycan, BM1-proteoglycan, have been identified in basement membranes. The chemistry, distribution, and possible functions of these substances will be discussed below.

### A. Neoplastic Basement Membranes As Models

Tumors have proven useful sources of basement membrane components. Pierce[20] characterized tumor basement membrane components and found that a parietal yolk sac (PYS) tumor produced a basement membrane-like matrix. This matrix was isolated free of cellular elements and used to immunize animals. The antibody produced in these animals reacted with the cells and matrix of the tumor and with basement membranes in a variety of normal tissues. These studies suggested that the components of tumor basement membranes resembled those in normal tissues.

Subsequently, Orkin et al.[21] found that a transplantable mouse tumor, the EHS sarcoma*, which had been described as a poorly differentiated chondrosarcoma, produced instead a matrix of basement membrane. In contrast to the PYS tumor, the EHS tumor will grow in a variety of mouse strains and is much less malignant to the host. The EHS tumor has proven to be a useful source of basement membrane constituents, including Type IV collagen, laminin, and a heparin sulfate (BM-1) proteoglycan.

* The designation of this tumor as a sarcoma is probably incorrect given the nature of its matrix, and we suggest that it be referred to as the EHS tumor.

**Table 1**
**COLLAGENOUS COMPONENTS OF BASEMENT MEMBRANES**

| Protein | Size (daltons) and subunit structure | Special features |
|---|---|---|
| Type IV collagen(10) | About 450,000 $[\alpha1(IV)]_3$, $[\alpha2(IV)]_3$ | Present in all basement membranes; helix contains noncollagenous segments |
| Type V collagen(15,16) | About 300,000 $[\alpha1(V)]_2$, $[\alpha2(V)]$, $[\alpha3(V)]_3$ (formerly αB, αA, αC) | Found also separate from basement membranes |
| 7-S collagen(17) | 360,000 (after reduction chains ranging from 25,000–150,000) | Collagenase-resistent fragment, disulfide-rich, cross-linking region of Type IV collagen |
| Intima collagen(16,18) | About 270,000 (after proteolysis), two to three chains of 40,000–50,000 | Localization not yet clarified; compact structure, disulfide-rich |

# II. COLLAGENOUS PROTEINS

## A. Type IV Collagens

All basement membranes contain a special collagen type, Type IV, which is considered to be the major structural component.[9] Type IV collagen shows structural differences that distinguish it from interstitial collagens which are reflected in a characteristic amino acid composition (see Table 2). In comparison to the interstitial collagens, it contains higher amounts of hydroxylated amino acids (4-hydroxyproline to proline ratio of 1.5:2; hydroxylysine to lysine >4; 3-hydroxyproline to 4-hydroxyproline ratio of 0.06:0.12) and a low content of alanine and arginine.[10] The glycine content is distinctly below one third, indicating the presence of noncollagenous segments. Most of the hydroxylysine residues are substituted by glucosyl-α(1:2)-galactosyl-β groups linked to the hydroxyl group.[22] Heteropolysaccharide side chains consisting of glucosamine, mannose, galactose, fucose, and sialic acid have also been identified as part of the Type IV collagen molecule, but their linkage is unknown.[23,24] Carbohydrate accounts for some 10% of the mass of Type IV collagen, a higher level than that found in most other collagens. A schematic model (see Figure 2) indicates that Type IV collagen consists of a large triple-helical domain and a few noncollagenous segments, a structure occasionally referred to as "procollagen-like".[4,10,25] However, this collagen is unique in that there are frequent interruptions of the triplet sequence Gly-X-Y within the triple-helical domain with glycine replaced by other amino acids.[26,27] A repetitive Gly-X-Y sequence is considered to be essential for maintaining the triple-helical conformation,[28] and such interruptions are not found in other collagens. So far, two genetically distinct constituent polypeptide chains, α1(IV) and α2(IV), and possibly a third chain, α3(IV), have been identified in this protein (see below). These chains may exist in various proportions in different triple-stranded molecules and thus give rise to three or four different species.

Properties of Type IV collagen have also been studied with tissue culture-derived material (see Figure 3). The form with the largest chains is obtained from cell and organ cultures and is referred to as Type IV procollagen. The Type IV procollagen isolated from cultures of amniotic epithelial cells[29] was found to have a composition resembling Type IV collagen from tissue sources (see Table 2). Electrophoresis of this material under reducing conditions shows two chains named proα1(IV) and proα2(IV) with apparent molecular weight of 185,000 and 175,000 daltons, respec-

## Table 2
## AMINO ACID COMPOSITION OF PURIFIED BASEMENT MEMBRANE PROTEINS (RESIDUES/1000)

| Amino acid | Procollagen Type IV[a] | Collagen Type IV[b] | Collagen Type IV pepsin resistant[b] | 7-S collagen (long form)[c] | Ascaris collagen Type IV[d] | Intima collagen[e] | Laminin[f] | Epithelial glycoprotein[g] | Heparin sulfate proteoglycan[h] |
|---|---|---|---|---|---|---|---|---|---|
| 3-Hyp | 7 | 6 | 7 | 3 | — | — | — | — | — |
| 4-Hyp | 93 | 104 | 127 | 115 | 126 | 65 | — | — | — |
| Asp | 54 | 53 | 50 | 50 | 46 | 78 | 107 | 99 | 81 |
| Thr | 30 | 34 | 27 | 26 | 14 | 16 | 57 | 64 | 68 |
| Ser | 49 | 57 | 53 | 29 | 19 | 27 | 76 | 77 | 84 |
| Glu | 91 | 102 | 93 | 91 | 82 | 104 | 120 | 90 | 154 |
| Pro | 69 | 65 | 61 | 77 | 87 | 92 | 52 | 57 | 81 |
| Gly | 269 | 281 | 319 | 326 | 314 | 318 | 91 | 81 | 106 |
| Ala | 49 | 44 | 30 | 25 | 58 | 41 | 75 | 79 | 70 |
| Cys | 13 | 7 | 6 | 18 | 8 | 20 | 44 | 21 | n.d. |
| Val | 36 | 29 | 28 | 30 | 19 | 21 | 48 | 57 | 63 |
| Met | 11 | 13 | 11 | 12 | 11 | 8 | 14 | 8 | 13 |
| Ile | 32 | 25 | 22 | 19 | 22 | 20 | 42 | 40 | 26 |
| Leu | 63 | 52 | 45 | 49 | 76 | 24 | 90 | 102 | 81 |
| Tyr | 15 | 9 | 5 | 10 | 10 | 18 | 27 | 38 | 24 |
| Phe | 33 | 29 | 29 | 25 | 17 | 15 | 31 | 47 | 32 |
| His | 13 | 11 | 8 | 8 | 5 | 3 | 24 | 26 | 32 |
| Hyl | 33 | 37 | 46 | 46 | 36 | 48 | 1 | n.d. | — |
| Lys | 20 | 11 | 5 | 7 | 17 | 18 | 51 | 68 | 29 |
| Arg | 37 | 33 | 29 | 34 | 38 | 64 | 50 | 46 | 54 |

[a] From amniotic fluid cells.[29]
[b] Acid-soluble form from mouse EHS tumor.[25]
[c] From human placenta.[17]
[d] From *Ascaris suum* intestinal basement membrane.[67]
[e] From aortic intima or human placenta.[16,18]
[f] From mouse EHS tumor.[93,107]
[g] From mouse kidney epithelium.[165]
[h] From mouse EHS tumor.[163]

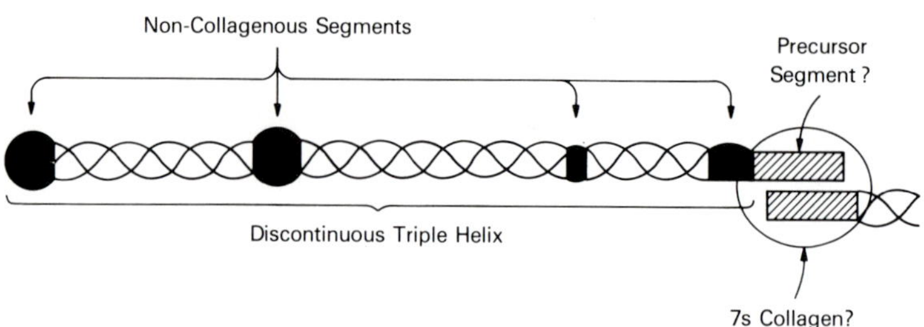


FIGURE 2. Schema of different structural domains in Type IV collagen. The order, location, and relative size of the various segments are arbitrarily assigned.

tively.[29–31] Peptide map analysis of both chains demonstrated that they are structurally distinct.[31–34] In culture, no conversion of Type IV procollagen to a smaller protein is observed, and there is some question whether Type IV procollagen undergoes conversion to a smaller protein (Type IV collagen) in vivo.

Soluble forms of Type IV collagen can be extracted with acidic solvents (usually dilute acetic acid) from the matrix of the EHS tumor[21,25] and from bovine lens capsule.[35,36] When electrophoresed, the reduced tumor collagen shows two chains $\alpha 1$(IV) and $\alpha 2$(IV), with apparent molecular weight of about 160,000 and 140,000.[25,37] The chains are not separated by molecular sieve chromatography, and their resolution on electrophoresis may be due to compositional differences more than to differences in molecular weight. Polypeptides similar in size to $\alpha 1$(IV) and $\alpha 2$(IV) are observed after solubilization of lens capsule basement membrane by limited pepsin digestion.[38] Ultrastructural studies suggested that the pepsin solubilized collagens contain globular structures at each end of the molecule.[39] Two collagenous polypeptides ($M_r$ = 164,000 to 187,000) were also found in renal basement membranes solubilized by reduction under denaturing conditions.[40,41] Such extracts contained larger collagenous species presumably formed from $\alpha 1$(IV)- and $\alpha 2$(IV)-chains cross-linked by lysine-derived aldehydes known to occur in Type IV collagen.[42]

As expected, the triple helix of Type IV collagen is digested by bacterial collagenase resulting in a mixture of small peptides.[4,21] Two peptides containing noncollagenous sequences have been isolated from such a digest of Type IV collagen (see Figure 4). These peptides ($M_r$ = 10,000 to 20,000) contained little or no hydroxyproline and hydroxylysine and were poor in glycine, but contained cystine. They presumably represent two of the larger noncollagenous segments of Type IV collagen (see Figure 2).

Pepsin and other proteases are used to dissolve otherwise insoluble collagen. Considerable heterogeneity was observed in the Type IV collagen solubilized with pepsin from glomeruli or kidney cortices,[43–47] lens capsule,[3,36,48–50] placenta,[51–54] blood vessels, muscle[55] and the EHS tumor,[25,56] with polypeptides that ranged from $M_r$ = 15,000 to 140,000 (see Figure 3). Distinct fragments were also obtained with a variety of neutral proteases including leucocyte elastase.[25,54,57–59] It is now clear that in contrast to other collagens, Type IV collagen contains nonhelical regions interrupting the helical domain which are sensitive to various proteases. The sensitivity of such regions to protease may be significant in the removal of basement membrane. The collagenases from vertebrate tissues which carry out the initial attack on interstitial collagens (Types I to III) do not degrade Type IV collagen.[60,61] However, another enzyme specific for Type IV collagen was isolated from an in-

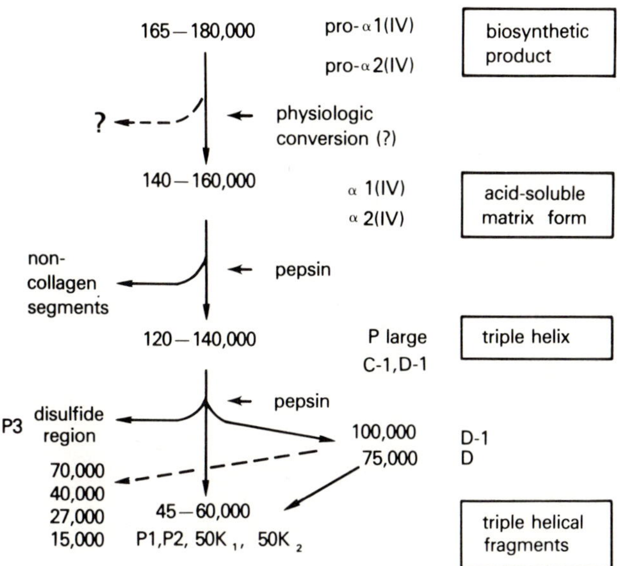


FIGURE 3. Relationship between two different constituent chains of Type IV collagen and fragments produced from them by proteolysis. The various fragments originate either from the α1(IV)-chain (P large, C-1, C,α, P1, P3, $50K_1$) or the α2(IV)-chain (P large, D-1, D,α′, P2, $50K_2$). The molecular weights indicated were determined by SDS electrophoresis or agarose chromatography and should be considered tentative.

vasive tumor,[62] suggesting that there is a specific pathway for the turnover of this collagen.

The isolation and characterization of various pepsin fragments has been recently carried out in several laboratories and will be summarized here and related to the α1(IV)- and α2(IV)-chains. While the peptides derived from Type IV collagen with pepsin are diverse, they separate when chromatographed on cation exchange resins into two sets of fragments. The more acidic fragments arise from the α1(IV)-chain and have a low arginine content (14 to 24 residues per 1000 amino acids), but higher 4-hydroxyproline and hydroxylysine. The more basic fragments in the pepsin digest arise from the α2(IV)-chain and have a higher arginine content (32 to 47 residues per 1000 amino acids) and tend to be more hydrophobic.[36,43–47,50,52,53,55,56]

A kinetic study has been made of the degradation of Type IV collagen prepared from the EHS tumor by pepsin[56] (summarized in Figures 3 and 5). The biosynthetic product is larger than the chains present in the molecules extracted from the tumor with acid.[30] Treatment of the acid-soluble matrix form with pepsin caused the release of small noncollagenous peptides presumably derived from either one or both ends of the molecule. The α2(IV)-chain is more rapidly degraded than the α1(IV)-chain[25,27,56] which may account for the recovery of only $\alpha1(IV)_3$ molecules in earlier studies.[3,48] Large triple-helical fragments with increased levels of glycine, hydroxyproline, and hydroxylysine (see Table 2) persisted longer. These were found to contain chains ($M_r$ = 130,000 to 140,000) linked by disulfide bonds. Cleavages in the central portion of the triple helix produce P2, P3B, and P3C from the α2(IV)-chain and P1 and P3A from the α1(IV)-chain. Quite similar cleavages occur in Type IV collagen from chick.[55]

Another pattern emerged from studies of pepsin fragments from Type IV collagens

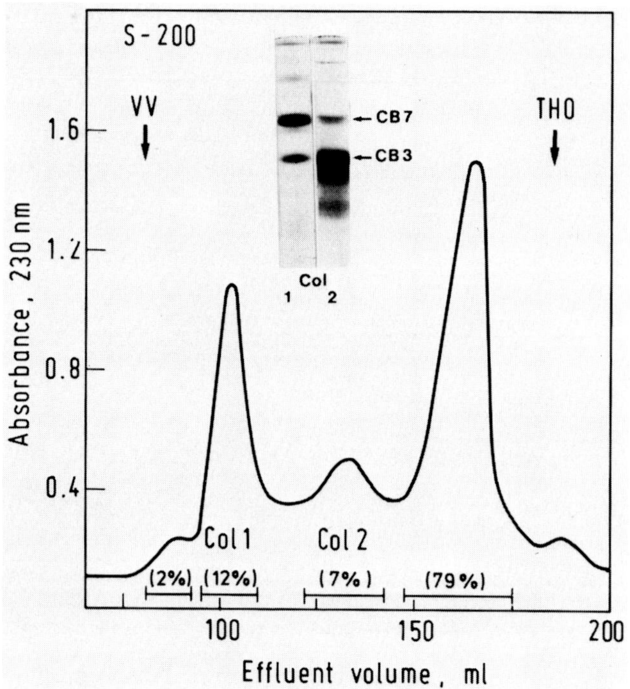


FIGURE 4. Separation of two noncollagenous segments (Col 1, Col 2) obtained by collagenase digestion of acid-soluble Type IV collagen (mouse tumor) on Sephacryl® S-200. The insert shows SDS electrophoresis pattern of the crude peptides indicating mobilities similar to CNBr peptides α1-CB7 and α1-CB3 of Type I collagen. Number in brackets denote the relative amount (in percent) of peptide material collected in four different pools (horizontal bars). The column (1.5 × 110 cm) was equilibrated in 0.2 *M* ammonium bicarbonate, pH 8.5. The void volume (VV) and total volume (THO) of the column are marked by arrows. (Timpl, R., unpublished. With permission.)

from human, bovine, and porcine tissues (see Figures 3 and 5).[36,44,46,47,53] Here, the α1(IV)-chain is initially cleaved to a C-1 fragment ($M_r$ = 110,000 to 140,000) still linked by disulfides. Further cleavages produce the C fragment ($M_r$ = 95,000) and the $50K_1$ fragment ($M_r$ = 50,000). The $50K_1$ fragment is very similar to the P1 fragment of mouse Type IV collagen. Cleavage of the putative α2(IV)-chain produces fragments designated D-1, D, and $50K_2$. The D fragment was separated into two smaller fragments with $M_r$ = 70,000 to 80,000 (D[75]) and with $M_r$ = 15,000 (D[15]) by reduction of a single disulfide bond.[36,46,47,53] It was also reported that reduction under nondenaturing conditions followed by a second pepsin treatment facilitated the release of the C and D(75) fragments,[48,52] suggesting that the disulfide bond(s) impeded the protease. A few more fragments, particularly disulfide-linked variants of C and D(75), have been reported,[46,51,53,54] but their precise nature remains to be established. It has also not been determined whether the release of smaller fragments proceeds by a continuous degradation of C-1 or D-1 from their N terminal end or results from a few central cleavages as found for mouse and chick Type IV collagen.

Cyanogen bromide peptide patterns have been used to establish the relationship between various pepsin fragments. It should be noted that part of the methionine in basement membrane collagen may be oxidized[63] and thereby resistant to cyanogen bromide unless reduced prior to CNBr digestion.[26,53] The separation and charac-

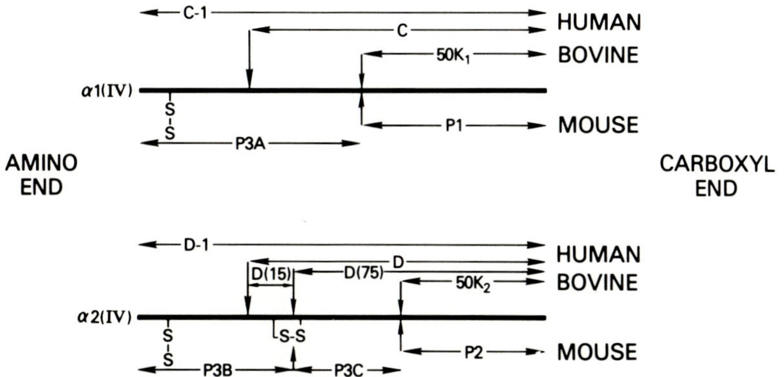


FIGURE 5. Tentative order of pepsin fragments in the triple-helical domain of α1(IV)- and α2(IV)-chain. The size of the various fragments produced from human placenta, bovine, or porcine kidney, and bovine lens capsule Type IV collagen[36,44,46,47,48,53] (C-1, D-1, C, D, $50K_1$, $50K_2$) or from mouse EHS tumor[56] Type IV collagen (P1, P2, P3A, P3B, P3C) is indicated by the horizontal arrows. Vertical arrows denote cleavage sites within the triple helix, S-S denote either inter- or intrachain disulfide bridges.

terization of CNBr peptides from C-1, C, and $50K_1$ fragments of bovine lens capsule,[50,64,65] the D(75) fragment of porcine kidney,[47] and of the P1 fragment[26] of mouse tumor has been accomplished. The observation of a common, short C terminal CNBr peptide in the C-1, C, $50K_1$, and P1 fragments[26,50,65] (see peptide P1CB7 in Figure 6) was crucial in deducing the order of the fragments in the α1(IV)-chain (see Figure 5). This illustrates the value of sequence data in assessing the relation between distinct peptides.

Partial aminoterminal amino acid sequences have been obtained for pepsin fragments and CNBr peptides of Type IV collagen from the EHS tumor,[26,56] human placenta,[27] and bovine lens capsule.[65] The data show sequences within the triple-helical region which deviate from the triplet structure Gly-X-Y (see Figure 6), indicating that the triple-helical conformation is discontinuous in these regions. The sequences interrupting the triplet structure include segments of three to eight amino acid residues at the N terminal end of fragments C, D(75), P1, and P2. Presumably these stretches of nontriplet sequence are even longer in the intact chain. Similar interruptions were also found within pepsin fragments or in certain CNBr peptides derived from them (compare peptides P2-CB and F fragment), indicating sequences which are apparently not readily cleaved by pepsin. The presence of nontriplet sequences may explain the characteristic pepsin cleavage pattern (see Figure 5), while interspecies variations in the sequence may account for the observed variations in proteolytic fragments. One interruption of the Gly-X-Y structure observed is based on a single deletion of glycine (peptide P1CB5 in Figure 6). This type of discontinuity resembles that found in the triple-helical domain of the complement component Clq, where it is associated with a bend in the triple-helical strands.[66] Preliminary sequence data on portions of mouse and human Type IV collagen[200] indicate that about 3% of the sequences lack the Gly-X-Y structure, suggesting that the Type IV collagens are more flexible than other collagens.

The chains in the triple-stranded molecules of Type IV collagen are connected by disulfide bonds. In the α1(IV)-chain, the disulfides are located at the N terminal end of the triple-helical domain (see Figure 5).[50] Possibly other interchain disulfide bridges exist in noncollagenous domains. The data also suggest the occurrence of an intrachain disulfide bond connecting the D(75) and D(15) fragment of the α2(IV)-

| | | |
|---|---|---|
| mouse | | |
| α1 (IV) | P1 fragment | His-Val-Asp-Met-Gly-Ser-Met-Gly-Gly-Gln-Gly-Gly-Asp-Gln-Gly-Gln-Gly-Gly- |
| | P1-CB5 | Gly-Pro-Hyp-Gly-Pro-Gln-Gly-Gln-Hyp-Gly-Leu-Hyp-Gly-Thr-Hyp-Gly-XXX-Pro-Val-Glu-Gly-Pro-Pro-Gly-Asp-Gln-Gly-Pro-Gln-Gly- |
| | P1-CB7 | Gly-Pro-Hyp-Gly-Thr-Hyp-Ser-Val-Asp-His-Gly-Ser-Phe-OH |
| α2 (IV) | P2 fragment | Pro-Gly-Met-Lys-Asp-Ile-XXX-Gly-Glu-XXX-Gly-Asp-Glu-Gly-Pro-Met-Gly-Leu- |
| | P2-CB3 | Gly-Ser-Hyp-Gly-Ile-Hyp-Gly-Ile-Gln-Lys-Ile-Ala-Val-Gln-Pro-Gly-Thr-Leu-Gly-Pro-Gln-Gly-Ser-Arg-Gly-Leu-Hyp-Gly-Ala- |
| α3 (IV) | P4 fragment | Phe-Asp-Met-Ser-Leu-Arg-Gly-XXX-Gly-Asp-Hyp-Gly-Phe-Hyp-Gly-Gln-Hyp-Gly-Met- |
| human | | |
| α1 (IV) | C fragment | Tyr-Phe-Asp-Leu-Gly-Leu-XXX-Gly-Asp-XXX- |
| α2 (IV) | (D75) fragment | Glu-Ala-Ile-Gly-Pro-Gly-Gly-Ile- |
| | E fragment | Pro-Gly-Met-XXX-Asp-Ile-XXX-Gly-Glu- |
| ? | F fragment | Tyr-Gly-Glu-Ile-Gly-Ala-Thr-Gly-Asp-Phe-Gly-Asp-Ile-Gly-Asp-Thr-Ile-Asn-Leu-Pro-Gly-Arg-Hyp-Gly-Leu-XXX-Gly-Glu-Arg-Gly- |

FIGURE 6. Selected aminoterminal sequences of pepsin fragments and CNBr peptides (CB) obtained from the triple-helical region of mouse EHS tumor[25,26] or human placenta[27] Type IV collagens. Segments which are underlined lack the Gly-X-Y structure. The complete sequence is shown for the C-terminal peptide P1-CB7 which was used to align various pepsin fragments as shown in Figure 5. The peptides from human material[27] have been in part renamed to those fragments shown in Figure 5. However, this does not necessarily mean that they are completely identical. XXX denotes unidentified residues.

chain[46,53] (see Figure 5). The occurrence of an intrachain disulfide bridge within the triple helix is not found in other collagenous proteins and could indicate the site of a bend in the molecule.

Some evidence exists for a third constituent polypeptide chain α3(IV). A disulfide-bonded fragment with the size of the C fragment of α1(IV)-chain was discovered in pepsin digests of the insoluble residue of the EHS tumor.[56] Further cleavage released a C-terminal fragment P4 ($M_r$ about 42,000). Sequence analyses on CNBr peptides derived from these fragments proved them (see Figure 6) to be different from fragments of similar size obtained from the α1(IV)- and α2(IV)-chains.[201] Invertebrate basement membranes contain collagenous protein with three chains with the amino acid composition characteristic for Type IV collagen (see Table 2). Such material was solubilized with pepsin from intestinal tissue of the helminth, *Ascaris suum*, and found after reduction to contain three polypeptide bands in electrophoresis ($M_r$ = 145,000 to 160,000).[67] Their relationship to the polypeptide chains of vertebrate Type IV collagen remains to be established.

The diversity of polypeptide chains suggests that several forms of Type IV collagen exist in the tissue. The proteins containing the α1(IV) and α2(IV) were partially resolved by sequential salt precipitation, suggesting that they occur in different triple-helical molecules, i.e., $\alpha1(N)_3$ and $\alpha2(N)_3$.[37] Similarly, it was possible to separate the pepsin fragments P1 and P2 as triple-helical molecules,[56] i.e., $(P1)_3$ and $(P2)_3$. The different sensitivity of proα1(IV)- and proα2(IV)-chains of the biosynthetic precursor and of α1(IV)- and α2(IV)-chains in human placenta and the EHS tumor towards pepsin is also compatible with two molecules.[27,30,34,56] A different situation was found for Type IV collagen from chick gizzard.[55] The fragments resembling P1 and P2 from the α1(IV)- and α2(IV)-chain of the mouse tumor remained associated in a ratio 2:1 during several purification steps, indicating the molecular composition $[\alpha 1(IV)]_2\alpha 2(IV)$. Such results point to the possibility of tissue as well as species differences in the chain composition of the collagens in different basement membranes.

In spite of the sequence peculiarities indicating breaks in the helix, Type IV collagen has a thermal stability equal to[25] or even slightly higher[68] than that of interstitial collagens. The presence or absence of disulfide bonds does not alter the thermal stability.[25] A preliminary ultracentrifugal analysis showed a molecular weight of 450,000 ± 50,000 daltons for acid-soluble mouse Type IV collagen consistent with molecules composed of three 150,000-dalton chains.[25] Higher aggregates linked by either disulfide bonds and/or lysine-derived cross-links exist in tissues.[4,37,40] The X-ray diffraction pattern of stretched lens capsule showed a weak reflection at 0.29 nm compatible with a triple-helical conformation, but indicating also a poorly ordered fibrillar array.[69]

Some studies have been carried out on fibers prepared from Type IV collagen. Type IV collagen does not precipitate as readily as other collagen types from solution under physiological conditions, and this difference has been exploited in separating the collagen types by thermal gelation.[70] Type IV collagen solubilized by limited proteolysis was precipitated when warmed, but the precipitate was composed of randomly oriented fibrils lacking periodicity.[38] Based on such observations, it was proposed that *in situ* the Type IV collagen molecules are arranged as a random meshwork of single molecules connected by covalent bonds.[39] In other studies, it was suggested that there was an aggregation of Type IV collagen into microfibrillar structures which interact with noncollagenous components to form the basement membrane.[4]

Attempts to form ordered lateral aggregates of the SLS type with intact Type IV collagen were negative.[25,35] However, following proteolysis of Type IV collagen where disulfide bonds had been reduced or to a lesser extent after protease treatment of unreduced Type IV collagen, it was possible to obtain ordered SLS crystallites.[25,39,52,71,72] The cross-striation pattern of these crystallites was distinct from those obtained with Types I, II, III, or V collagen. Since the cross-striation pattern is a direct reflection of the amino acid sequence on the triple-helical domain, this indicates that there are numerous structural differences between these proteins.

The length of the SLS crystallites also reflects the length of the triple helix. With Type IV collagens, end-to-end aggregates of molecules (referred to as fibrous segment long spacing crystallites[39]) are usually observed, and the length of the molecules can only be estimated from the distance between mirror planes. A length of about 330 to 350 nm was observed for trypsin-treated Type IV collagen from the EHS tumor.[25] Shorter segments of about 300 nm were obtained from material solubilized by low concentrations of pepsin from bovine lens capsule.[71] These crystallites exhibited a dumbbell structure attributed to the presence of globular domains at both ends of the molecules which fold back over the triple helix.[39,72] Shorter segments of 150 to 260 nm were obtained after a more rigorous proteolytic treatment[25,52,71] in accordance with the chemical data discussed before. The size of the polypeptides comprising the triple helix of Type IV collagen was estimated by chemical analyses to be in the range 110,000 to 145,000.[50,53,54,56] Considering a content of about 10% carbohydrate, these estimates indicated that the triple helix is at least as long as found for interstitial collagens (290 to 300 nm), but may be as much as 20 to 30% longer.[25,54] In fact, rotary shadowing visualizes Type IV collagen obtained by limited pepsin digestion as strands with a total length of 400 nm, including the 7-S collagen domain (see below).[73] This indicates that the main triple helix has a length of about 340 nm. Rather sharp bends are observed in the individual strands which is compatible with an increased flexibility of the triple helix in Type IV collagen.

### B. Type V Collagens

Type V collagen was discovered in pepsin digests of placental membranes[15,16] as well as several other tissues.[16] Type V collagen was observed to be more soluble than other collagens particularly at high concentrations of NaCl (3 to 3.5 *M*) at neutral pH, conditions that readily precipitate the interstitial collagens. In addition, the protein does not aggregate under conditions that cause the heat gelation of Types I and III collagen.[70] Type V collagen shares some features of interstitial and of basement membrane collagens. For example, its amino acid composition in large part resembles interstitial collagens except for a high ratio of hydroxylysine to lysine and a low content of alanine.[15,16,74–77] Similar to interstitial collagens, the hydroxylysines are only partially glycosylated with glucosylgalactose or galactosyl groups.[16,75,77,78]

Some progress has been made in determining the histological distribution of Type V collagen with antibodies to the protein. These studies, based on both immunofluorescence[79,80] and immunoelectronmicroscopy,[81] indicate that Type V collagen is located in some, but not all, basement membranes. In contrast, Type IV collagen is present in all basement membranes. In support of the presence of Type V collagen in basement membrane is its isolation from Descemet's membrane of the cornea.[75] It is particularly abundant in muscle.[80] However, Type V collagen also occurs in tissues which lack morphological correlates of basement membranes, including corneal stroma,[75,82] tendons,[76,83] hyaline cartilage,[74,75,84] and synovial tissue.[78,85] These studies indicate that Type V collagen is not limited to basement membranes.

So far three chains, α1(V), α2(V), and α3(V) (formerly named αB, αA, and αC),[9] have been obtained from Type V collagen. These chains are separable on CM-cellulose[15,16,76,77] DEAE-cellulose,[74] phosphocellulose,[77,86] or hydroxylapatite.[75] They are similar in size, based on their behavior on molecular sieve chromatography, to the α-chains of interstitial collagens except that the α1(V)-chains appear to be some 10% larger.[76,86] The chains of Type V collagen show a slightly lower electrophoretic mobility compared to the chains of Type I collagen.[15,74,75,77,78] This may be due to a higher carbohydrate content or anomalous behavior, since their SLS crystallites are similar in length to those formed by interstitial collagen. Indeed the SLS crystallites of Type V collagen resemble those of the interstitial collagens more than they do basement membrane collagen.[74,75] However, Type V collagen is not cleaved by the animal collagenases that degrade the interstitial and Type IV collagens.[62,75,77] Together the data indicate that Type V collagen represents a distinct type in the collagen family.

The α1(V)-, α2(V),[15,16,74–77,86] and α3(V)-chains[77] differ from each other in composition and in the pattern of peptides produced by digestion with CNBr[15,76–78,86] or mast cell chymase[77] as expected for genetically distinct chains. One report suggests that α1(V)-, α2(V)-, and α3(V)-chains are derived from a single precursor chain,[85] but this seems unlikely given the other data.

Circular dichroism[74] and optical rotation[86] measurements indicate that Type V collagen is triple helical and has a thermal stability comparable to other collagenous proteins. The constituent polypeptide chains of Type V collagen are not connected by disulfide bonds as found in Type IV collagen, although it should be noted that the protease used to solubilize the protein could have cleaved off such portions (see below). Characterization of the CNBr peptides from the α1(V)-chain indicates that it comprises, when isolated by pepsin, some 1018 amino acid residues and is similar to the size of the chains of interstitial collagens.[87] It is likely that the α1(V)-chain is not intercepted by nonhelical segments, since it contains one third glycine and is not susceptible to degradation by pepsin in native form.

The chain composition of Type V collagen is still a matter of controversy. Some studies[15,74–76] indicate a ratio of α1(V)- to α2(V)-chains of 2:1, while a 2:1 ratio was observed in other studies.[78,83,86] Renaturation experiments[74] showed a greater thermal stability for the products $[\alpha 1(V)]_2\alpha 2(V)$ and $[\alpha 1(V)]_3$ than for $[\alpha 2(V)]_3$. Together with the compositional data discussed above, this was interpreted as indicating that Type V collagen is mainly composed of $[\alpha 1(V)]_2\alpha 2(V)$ molecules. Other data[86] showed a biphasic melting curve for placental Type V collagen, suggesting the presence of molecules of different stability and an enrichment of $[\alpha 1(V)]_3$ molecules after partial denaturation. Molecules containing α3(V)-chains are found only in certain tissues,[77,85] suggesting that they can be formed separately from the other chains of Type V collagen. Thus, the chains of Type V collagen may form different molecular species and show considerable heterogeneity.

The molecular structure of the molecule as deposited in the matrix is not known, nor have quarter-staggered fibrils of Type V collagen been identified in vivo or in vitro. Small amounts of Type V collagen chains extracted from embryonic tendons by acetic acid were found to be similar in size to those in protease digests.[83] Recent biosynthetic studies[32] on placenta indicate that Type V collagen contains disulfide-bonded chains slightly larger in size (difference in $M_r$ = 10,000 to 20,000) than α1(V)- and α2(V)-chains. It is possible that they represent either the intact molecule or a precursor form. Even larger chains were detected in a study of chick tissues.[88] It is still uncertain whether processing of Type V collagens resembles that of interstitial collagens by release of amino and carboxyl propeptides or follows the pattern suggested for Type IV collagen.

### C. 7-S Collagen

7-S Collagen was discovered in a collagenase (bacterial) digest of the EHS tumor matrix.[59] Surprisingly, although isolated with collagenase, it was found to have a composition resembling Type IV collagen (see Table 2).[17] Chromatographic and physical studies indicated that it was a relatively homogeneous species, and it received its name from its sedimentation constant. In the initial study, it could not be excluded that 7-S collagen was a new collagen type with an unusually compact structure. However, as discussed below, its appearance and the small size of the polypeptides obtained after reduction suggest that 7-S collagen is derived from a second triple-helical domain of Type IV collagen involved in cross-linking.

Solubilization of 7-S collagen from tissues requires proteolytic treatment, but small amounts are detected with immunological methods in acid extracts of tissues[17,59] and in cell culture medium.[89] A large form of 7-S collagen ($M_r$ = 360,000) is obtained when the digestion with bacterial collagenase is carried out at 20°C. A shorter variant with $M_r$ = 225,000 is produced from the long form by collagenase digestion at 37°C. Electronmicroscopy visualized the short form as compact rod-like particles with a length of 29 nm (see Figure 7). The long form showed in addition, four thinner strands (length 30 nm) extending from both ends of the central core in a symmetric fashion.[73] Hydrodynamic data agree with these pictures indicating dimensions of 3.4 × 49 nm for the short form and 3.1 × 95 nm for the long form.[17] These data suggest that 7-S collagen consists of four short triple-helical strands arranged in a parallel or antiparallel alignment (see Figure 8).

Both forms of 7-S collagen have a highly stable triple helical domain which melts

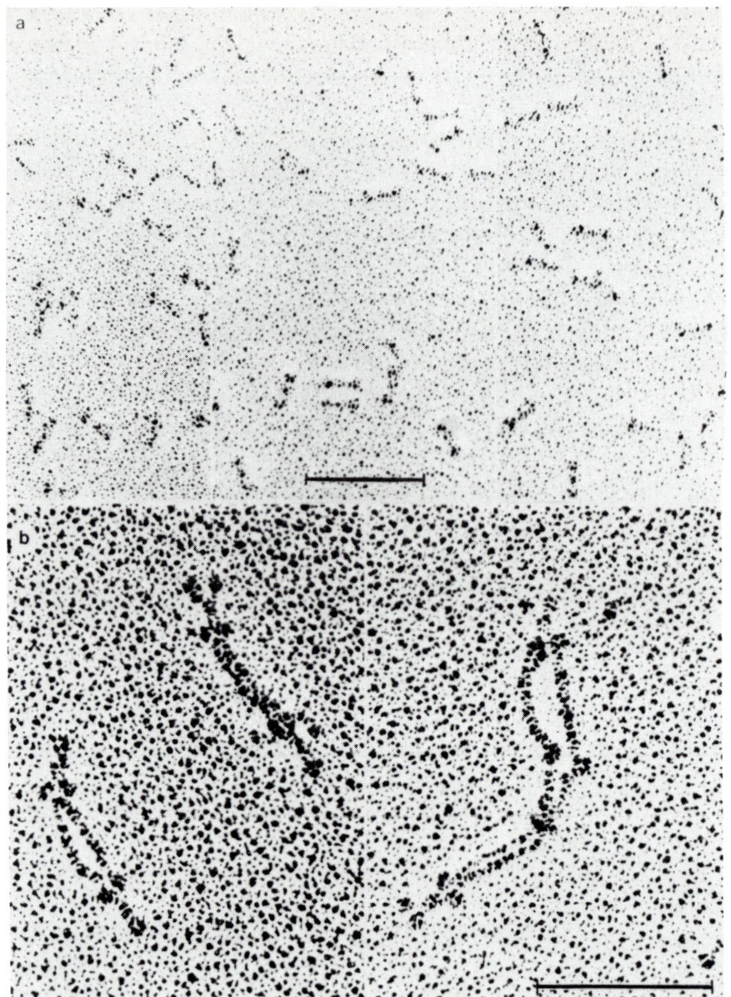


FIGURE 7. Electronmicrographs obtained by the rotary shadowing technique of the short form of 7-S collagen from human placenta[17] (A) and of intima-type of collagen from human placenta (B). The bars indicate a length of 100 nm. Note in (A) the rather uniform appearance of short compact rods of 28 nm. The pattern in (B) shows two apparently dimeric molecules in which two triple-helical strands are aligned to each other and connected via globular nodules (left-hand side) and an end-to-end aggregate of these structures (right-hand side). (We are indebted to Dr. H. Furthmayr for these photographs.)

at 70°C compared to 38°C to 41°C for other collagens.[71] The additional domains of the long form extend as single triple helices on both sides of the more compact domain and show a $T_m$ of only 48°C. Reduction of disulfide bonds in 7-S collagen under nondenaturing conditions produces a single melting profile with $T_m$ = 48°C. This indicates that cysteine (40 to 45 residues per 1000 amino acids) are located within the more stable domain. Both domains also showed other differences in chemical and immunological properties which in part may be due to short noncollagenous structural elements. Common to both domains is a high carbohydrate content (20 to 25%), including glucosylgalactosyl groups bound to hydroxylysine and heteropolysaccharide chains containing mainly glucosamine and mannose.[17]

A heterogeneous mixture of peptides from 15,000 to 150,000 daltons and higher

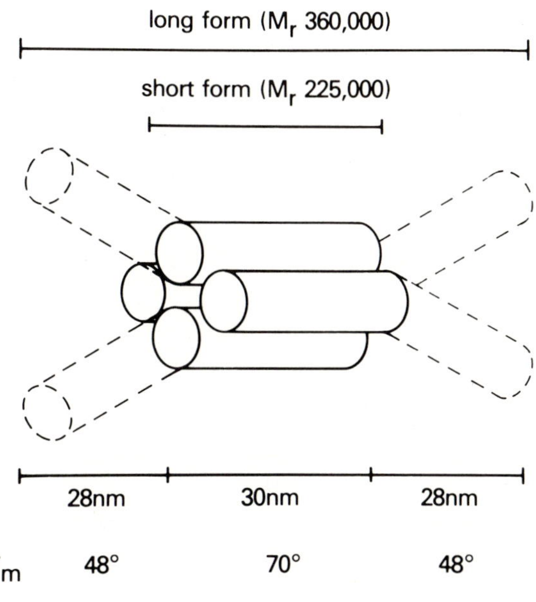


FIGURE 8. Tentative model of 7-S collagen assuming the lateral association of four triple-helical domains (cylinders) connected to each other by disulfide bonds and other cross-links. The full line represents the structures in the short form, the full- and broken-line structures in the long form of 7-S collagen as estimated by electromicroscopy.[73] The various domains within a single triple helix differ also in thermal stability as indicated by a difference in $T_m$.[17]

was obtained from the reduced proteins on electrophoresis.[17] It is probable that the extent of proteolytic cleavage during the preparation of 7-S collagen is variable. However, this heterogeneity is not due solely to the use of collagenase, since 7-S collagen with a similar heterogeneity was prepared by extensive pepsin digestion alone.[73]

Type IV collagen solubilized by limited proteolysis showed, after rotary shadowing, abundant numbers of spider-like structures. The center of this structure had the appearance and dimensions of the short form of 7-S collagen, but the four protruding extensions were much longer (about 370 nm) than those of the long form of 7-S collagen.[73] Acid-extracted Type IV collagen showed much less of these spider-like structures in agreement with immunological data on a low content of 7-S collagen.[59] This material, however, possesses globular nodules at the end of the strands which is not found in the pepsin-solubilized material.[202] Apparently, both materials contained incomplete Type IV collagen molecules which differed in the segments they had lost. Taken together the data suggest that 7-S collagen is a unique structure possibly generated by the cross-linking of segments of Type IV collagen (or Type IV procollagen). Type IV collagen molecules may associate through only one or two small domains which represent the site of disulfide- and lysine-derived crosslinks.[202]

### D. Intima Collagen

Still another distinct collagenous protein, rich in disulfides, was found in pepsin digests of the intima of the aorta.[16] Intima collagen was also obtained from human and bovine placenta[17,18,90] and human liver.[91] This protein was readily separated from other collagens because of a high solubility in NaCl solutions at acidic and neutral pH. However, it precipitates when dialyzed against dilute phosphate buffer at neutral pH. It is not an abundant collagen, and placenta, for example, has only 3% as much intima as Type IV collagen.[18]

Intima collagen has a high hydroxylysine content (see Table 2), as well as cystine and carbohydrate, including glucosamine, and in these regards resembles 7-S collagen and Type IV collagen. However, it has a high arginine content, lacks 3-hydroxyproline, and has a 4-hydroxyproline to proline ratio of less than one. Glycine comprises less than one third of the residues, indicating the presence of nonhelical domains in the protein.[17,18,90] Two chains were obtained after reduction and denaturation which differed from each other in composition and CNBr peptide pattern.[18,90] The composition of the CNBr peptides from the more acidic chain indicate that noncollagenous sequences occur throughout the chain.[18]

Intima collagen is resistant to digestion by bacterial collagenase and shows a biphasic thermal melting profile of the triple helix similar to that described for the long form of 7-S collagen.[90] However, its composition as well as immunochemical properties clearly distinguish it from 7-S collagen.[17] Further, its appearance is quite distinct (see Figure 7). Rotary shadowing visualized intima collagen as two to four strands 130 to 150 nm in length interrupted by at least two globular nodules.[203] After reduction under nondenaturing conditions, the globular nodules disappeared, and most of the strands became disconnected. The reduced material has a molecular weight of about 160,000 daltons, indicating that it is a single triple helix composed of three polypeptides with $M_r$ = 50,000 to 60,000. Prior to reduction, dimeric and tetrameric components were observed in the electron microscope (see Figure 7). The origin of intima collagen is not known. Preliminary studies show that antibodies to intima collagen do not react with basement membranes.[90] It is most likely to be a fragment of a new collagenous protein.

## III. NONCOLLAGENOUS PROTEINS

Basement membranes also contain significant amounts of noncollagenous glycoproteins which presumably account for their positive periodate-Schiff reaction,[19] as well as their ability to bind lectins.[92] Basement membranes from a variety of tissues (e.g., kidney, cornea, lens capsule) were found to contain between 10 and 50% noncollagenous proteins[4,10] estimated from the amount of hydroxyproline in these matrices, while the EHS tumor matrix contains more than 80% noncollagenous glycoproteins.[93] It is not yet clear whether such large differences in composition are real or due to differences in the isolation of the basement membranes. A number of proteins have been reported and the more completely characterized will be discussed here (see Table 3).

### A. Laminin

Laminin is a large basement membrane-specific glycoprotein. Its identification was facilitated by using the EHS tumor in which laminin comprises almost one half of the matrix proteins. Most of the laminin in the tumor can be extracted in neutral buffers of moderate ionic strength.[93] This allows its isolation in native form and with high biological activity. Alternatively, it can be isolated from the matrix produced by cultured endodermal[94–96] and teratocarcinoma cells.[97–100] In the latter case, it was isolated in native form after collagenase treatment.[95,99] Its subunits were isolated by electrophoresis after reduction and denaturation.[95,100] Large fragments of laminin are present in pepsin digests of placenta and kidney (see below). Immunohistology indicates that laminin is an abundant component of all basement membranes.[101,102]

**Table 3**
**NONCOLLAGENOUS COMPONENTS OF BASEMENT MEMBRANES**

| Protein | Size and subunit structure | Special remarks |
|---|---|---|
| Laminin(93) | About 800,000–1,000,000, containing 200,000 and 400,000 chains | Abundant component; asymmetric and elongated shape; cell adhesion protein |
| Fibronectin(110) | 450,000, containing two identical chains | Abundant in embryonic, but not adult basement membranes; cell adhesion protein |
| Heparan sulfate proteoglycan(163) | About 500,000–1,000,000, 6–12 glycosaminoglycan chains ($M_r$ = 70,000) | Minor, but ubiquitous, component; controls permeability of basement membranes |
| Epithelial glycoprotein(165) | Large aggregate composed of 34,000 chains | Restricted to epithelial basement membranes |
| Bullous pemphigoid antigen(172) | ? not known | Restricted to epidermal basement membrane |
| Goodpasture antigen(167) | ? not known | Restricted to kidney and lung basement membranes |

The amino acid composition of laminin is not sufficiently unique to be characteristic (see Table 2), but does differ, for example, from that of fibronectin.[93] Laminin contains about 12 to 15% carbohydrate with a distinct amount (4 to 6%) of sialic acid.[100,103] While the native protein migrates in electrophoresis as a narrow band, reduction of disulfide bonds produces two broad, faster migrating bands, with $M_r$ = 200 to 220,000 and 400 to 440,000 (see Figure 9). The bands often appear as doublets, particularly with biosynthetically labeled material.[31,94,98,104] Separation and chemical characterization of the $M_r$ = 200 to 220,000 and 400 to 440,000 peptide chains has been difficult due to aggregation of the chains after reduction and denaturation. Amino acid analyses of chains separated by electrophoresis have given conflicting results with one study showing distinct differences[100] and the other similarities[95] in composition. Their size suggests a monomer-dimer or precursor-product relationships.[93] However, as discussed below, the shape of the intact proteins makes it more likely that these are chemically distinct chains which are subunits of the intact molecule.

Ultracentrifugal analyses indicate that laminin has a molecular weight in the range 800,000 to 1,000,000 daltons, both in neutral buffer and under dissociating conditions (6 *M* guanidine).[103,105] However, due to the instantaneous formation of aggregates in these studies, precise measurements were not obtained. A sedimentation constant of $S^o_{20w}$ = 11.5S was obtained which is compatible with an elongated shape for the molecule. In electron microscopy, laminin is visualized as a cross with asymmetric arms (see Figure 10). Similar pictures were obtained by rotary shadowing and negative staining and also by electronmicroscopy of unstained molecules.[105] Using the latter technique, the mass of laminin was estimated to be about 900,000 daltons. A comprehensive analysis of laminin electron micrographs is consistent with the structural model for the protein shown in Figure 11. Laminin has a long arm (77 nm) and three short arms (36 nm). The diameter of these arms is about 2 nm, with larger globules (diameter 5 to 7 nm) at each end and in the central region of the shorter arms. The data also indicated that the arms are fixed in some rigid position rather than being entirely flexible.

Circular dichroism studies demonstrated about 30% $\alpha$ helix in laminin, together with some $\beta$ structure and aperiodic structural elements.[103,106] The $\alpha$ helix showed

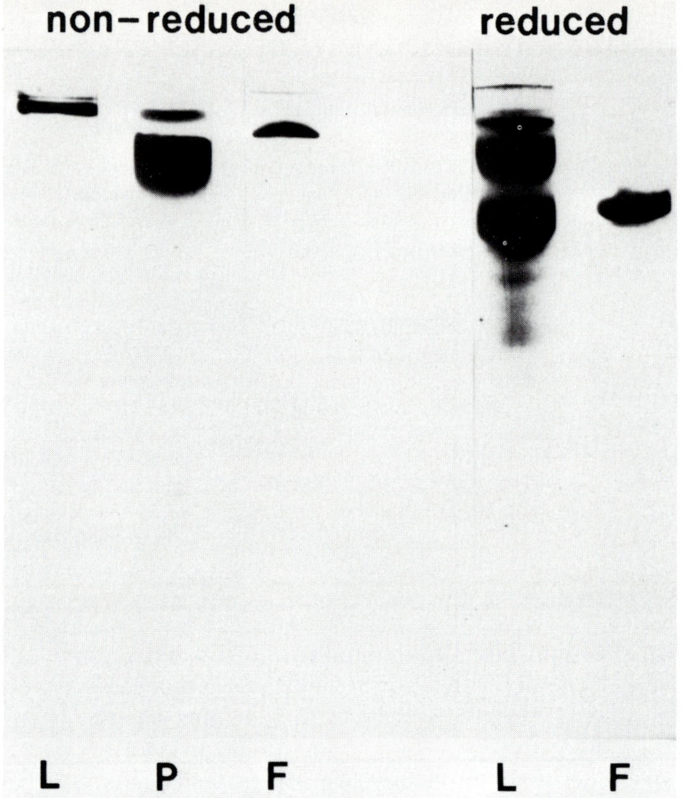


FIGURE 9. Sodium dodecylsulfate electrophoresis of laminin (L), laminin fragment P1 (P), and plasma fibronectin (F). Samples were analyzed in nonreduced form and after reduction with 2-mercaptoethanol; 3.5% gels.

a sharp thermal transition at 58°C, and this region was particularly susceptible to proteolytic degradation. Plasmin, however, only degraded it after denaturation.[106]

The proposed multidomain structure of laminin as suggested by electronmicroscopy was supported by proteolytic cleavage. Pepsin releases a large disulfide-rich (150 Cys per 1000) fragment P1 ($M_r$ = 290,000) (see Figure 9) and another disulfide-containing peptide P2 ($M_r$ = 45,000) together with a number of smaller fragments.[107] Fragment P1 was localized by electron microscopy to the region of the molecule where the arms intersect.[105] Similar fragments were prepared from pepsin digests of human placenta and kidney.[108] Cleavage with neutral proteases allowed the identification of a further fragment similar in size to P2 which showed affinity for heparin as well as a different fragment with $M_r$ = 70,000 to 90,000.[106]

Recent data indicate that laminin is a multifunctional protein and the correlation of specific functions with distinct domains of the molecule is under study. Laminin, perhaps through a small globular unit,[106] has the ability to interact with heparin and heparin sulfate.[99] It can also interact with Type IV collagen and cells.[109] Recent studies indicate that laminin is an attachment protein able to link epithelial and endothelial cells to basement membranes.[109] These studies have shown that laminin is active in cell attachment at levels of 1 to 5 μg/mℓ and binds to Type IV, but not to other collagens. Further, it is specific for epithelial and endothelial cells and shows no attachment activity with fibroblasts or chondrocytes.

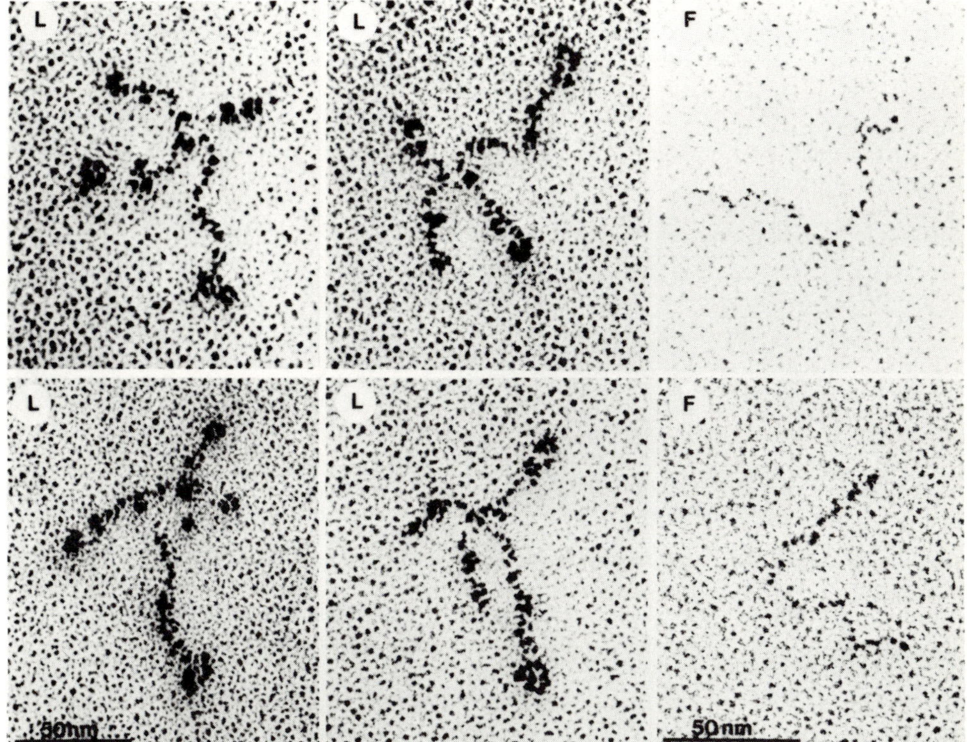


FIGURE 10. Visualization of single molecules of laminin (L) and of plasma fibronectin (F) after rotary shadowing in the electron microscope. The bar indicates a length of 50 nm. Photographs were supplied by Dr. H. Furthmayr.

**B. Fibronectin**

Since fibronectin is not restricted to or a constant component of basement membrane, only certain structural and functional aspects will be reviewed here. More comprehensive treatments of fibronectin are present in several recent reviews.[110–116]

Fibronectin was identified as a major product of cultured fibroblasts[117] and later shown to be similar or identical to a serum protein,[118,119] described earlier. Further studies have demonstrated that fibronectin is produced by a large variety of cells,[110–116] including endothelial and smooth muscle cells and even some epithelial cells. However, other studies suggest that the fibronectin produced by all these cells is not identical.[104,120–122] Immunohistology studies have shown that fibronectin is produced early in development[123] and is associated with most embryonic basement membranes. However, following embryogenesis fibronectin is found in some, but not other, basement membranes.[124] It has been possible to localize fibronectin at the ultrastructural level to basement membrane in the mesangium[125] and to discrete areas of the glomerular lamina rara.[126]

Fibronectin is present both in the matrix and on the plasma membrane of cultured cells.[127] Fibronectin has an affinity for collagen and is often distributed along collagen fibers.[127a] Fibronectin is also found in fibers which lack a distinctive ultrastructural appearance which are probably formed by self assembly of fibronectin as demonstrated in vitro with purified fibronectin.[128] Such fibrillar deposits become cross-linked via disulfide bonds[115] presumably involving the free sulfhydryls in the molecule. γ-Glutamyl cross-links are formed in fibronectin from glutamyl and lysyl

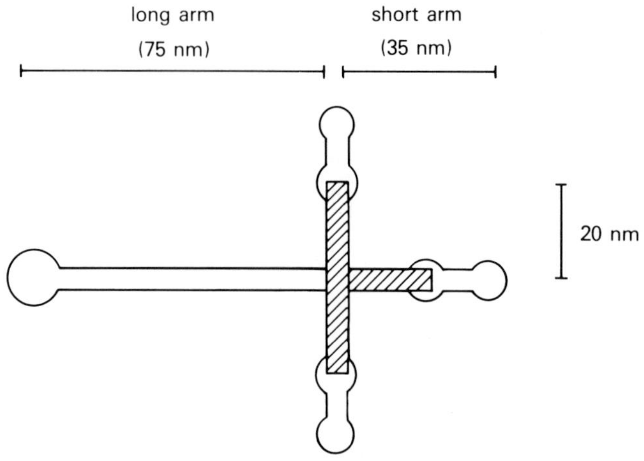


FIGURE 11. Structural model of laminin based on electron micrographs obtained from single molecules by rotary shadowing, negative staining, and scanning transmission electron microscopy.[105] The hatched area denotes a possible position of the large fragment P1 obtained from laminin by pepsin digestion.

residues by transglutaminase. Such cross-links can occur between fibronectin molecules, between fibronectin and fibrin, and between fibronectin and collagen.[129] The transglutaminase cross-linking site and the portion of fibronectin that binds to collagen have been localized to adjacent regions of the molecule (see Figure 12) which explains why α1-CB7 ($M_r$ = 25,000), a peptide from collagen, is able to cross-link to fibronectin after binding.[129]

Most chemical studies have been carried out with the plasma form of fibronectin. Fibronectin from the matrix or cell surface may be slightly different from the plasma form, particularly with respect to its hydrophobic nature, tendency to aggregate, and protease sensitivity.[121,129a,130] Fibronectin has a molecular weight of 450,000 daltons[131] and is thought to be composed of two identical polypeptide chains, since a single N terminal sequence PCA-Ala-Glx-Glx-Met-Val has been obtained from human plasma fibronectin.[131,132] The chains of fibronectin are connected to each other by disulfide bridges (see Figure 12) located close to their carboxyterminal ends.[133] Some separation of the chains is observed on electrophoresis, but this is usually ascribed to heterogeneity of carbohydrate side chains.[112] Each chain of plasma fibronectin contains three to four carbohydrate side chains[134] which are linked N-glycosidically to asparagine and consist of glucosamine, mannose, galactose, and sialic acid. Cellular forms of fibronectin may also contain other additional carbohydrate including fucose and galactosamine.[135]

The circular dichroism spectrum of fibronectin is distinct from that of laminin and indicates that the protein lacks α helix and β structure.[131,136] Ultracentrifugal analysis demonstrated a high frictional ratio of 1.7, indicating that the molecule has an elongated shape which increases further after denaturation.[136] Fibronectin is visualized by electron microscopy as a two stranded molecule with a length of each arm of 61 nm (see Figure 10). These arms are joined at one end in a rather rigid angle of 70° ± 20°C. The strands also showed several distinct bends, indicating that there are several rigid domains in the molecule connected by flexible segments.[105] As dis-

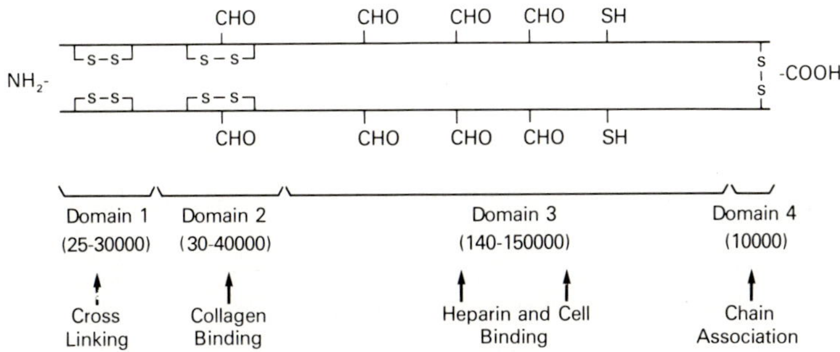


DOMAIN MODEL OF FIBRONECTIN

FIGURE 12. Multidomain model of fibronectin and localization of different binding activities. The domains were arbitrarily numbered, and their approximate molecular weights are indicated in brackets. Some special chemical features, such as amino ($NH_2$) and carboxyl (COOH) end of the peptide chains, disulfide bridges (S-S), free cysteine groups (SH), and carbohydrate side chains (CHO), are indicated. The model is based on studies of Wagner and Hynes,[144] McDonald and Kelly,[145] Yamada et al.,[146] Balian et al.,[147] Furie et al.,[148] and Mosher et al.,[149] among others.

cussed below, various activities of fibronectin are associated with different portions of fibronectin.

Proteolytic or chemical cleavage of fibronectin degrades the molecule producing a variety of peptides. Bonds near the carboxyl terminus of the molecule are particularly susceptible to cleavage by thrombin,[137] plasmin,[138] and trypsin.[139,140] Brief incubation with these enzymes removes the interchain disulfides and releases near full length chains. A more extensive digestion occurs with time using these or other neutral proteases.[139–149] A variety of enzymatic fragments have been isolated and tested for activity and have also been mapped to positions along the chain (see Figure 12). Peptides with intrachain disulfide bonds are located mainly in the aminoterminal regions, while carbohydrate is present predominently in the central portions. Fibronectin contains one or two free thiol groups at the carboxyl terminus, while the region that binds collagen is nearer to the amino end. In general, these studies, as well as physical[136] and electromicroscopic[105] data, indicate that the chains of fibronectin contain a number of distinct domains joined by exposed segments.

Fibronectin binds to a variety of other macromolecules. Fibronectin binds to several types of collagens,[150–152] particularly to denatured collagen. One of the major binding sites for fibronectin on the collagen molecule has a hydrophobic nature and resides in cyanogen bromide peptide α1-CB7 close to or including the cleavage site for animal collagenase.[151,153] Fibronectin also binds several glycosaminoglycans, particularly heparin.[154,155] The affinity constant is in the range $10^7 - 10^8$ mol/$\ell^{-1}$, and different sites for binding heparin and hyaluronic acid were identified.[146] Another important function was found in the interaction with cell membranes, possibly via interaction with certain gangliosides.[156]

## C. Proteoglycans

The first evidence for the presence of glycosaminoglycans in basement membranes came from biosynthetic studies on embryonic tissues.[157,158] Subsequently, polyanionic sites in native basement membranes were observed by ruthenium red staining in the glomerulus as discrete regular lattices with a distance of about 60 nm.[157,159,159a]

The anionic sites disappeared after treatment with heparitinase or heparinase[160] which also increased the permeability of the kidney and the basement membrane of the glomerulus. Heparin sulfate and hyaluronic acid were isolated after proteolytic digestion from purified glomerular basement membranes,[161,162] where they accounted for about 1% of the dry weight of the material.

More recently a heparin sulfate-containing proteoglycan (BM-1 proteoglycan) was extracted and purified from the EHS tumor matrix.[163] This material ($M_r$ = 500,000 to 1,000,000) consists of equal amounts of protein and heparin sulfate. The amino acid composition (see Table 2) of the protein core is not very exceptional except for a high glutamic acid content, but does differ from that observed with other basement membrane proteins. The heparin sulfate chains, when released by proteolysis from the proteoglycan, had an average molecular weight of 70,000 daltons, indicating that the intact molecule contains 6 to 12 of these chains. Antibodies prepared against the proteoglycan reacted mainly with the protein core and stained in immunofluorescent tests authentic basement membranes of skin, kidney, and cornea. The heparin sulfate-containing proteoglycan, which is larger in size than similar materials obtained from other sources such as liver, represents a common component of basement membranes. Other proteoglycans (BM-2)[163] may also be present but have not yet been characterized.

**D. Proteins with Restricted Distribution**

Basement membranes in different anatomical sites and/or produced by different cells do not exhibit the same ultrastructure, suggesting quantitative or even qualitative compositional differences. A restricted distribution was, for example, observed for fibronectin and may even exist for laminin since laminin was not detected in the perisinusoidal region of liver parenchyma which contains Type IV collagen.[95,164] Further, as discussed below, certain components appear to be present only in certain basement membranes.

A glycoprotein, specific for epithelial basement membranes, was extracted with neutral buffer from kidneys[165] and was also obtained from teratocarcinoma cell culture medium.[166] It consists of a large aggregate of polypeptide chains, $M_r$ = 34,000, linked by disulfide bonds. The amino acid composition of this protein is different from that of laminin and BM-1 proteoglycan (see Table 2). It contains about 13% carbohydrate, including glucose. Antisera against the protein react with the basement membranes in renal glomeruli and tubules, but not with the basement membrane in blood vessels. Its function is not yet known.

Further evidence for the restricted occurrence of some basement membrane proteins arose from studies with certain autoimmune diseases. Serum from patients with the Goodpasture syndrome[167] contains antibody which reacts exclusively with renal and alveolar basement membranes, but not the epidermal-dermal basement membranes. Patients with this syndrome present with lung hemorrhages and renal failure caused by nephritis. The nature of the Goodpasture antigen is still controversial. Data have been presented which indicate that it is either a set of noncollagenous proteins[168] or the carbohydrate region of basement membrane collagen.[169] It is apparently not identical with laminin or Type IV collagen[170] as expected from the restricted distribution of the antigen. In contrast, patients with bullous pemphigoid,[171] a blistering disease of the skin, have circulating antibody that reacts primarily with epidermal basement membranes. The latter antigen has been extracted from normal human skin and was reported to be composed of small polypeptide chains.[172]

It is likely that our knowledge about the total number of matrix components is still incomplete. In fact, a considerable variety of proteins or polypeptide chains are obtained by extracting basement membranes under denaturing or proteolytic conditions.[4,5,11,40,41,63,168,173–175] The nature of these components still remains to be determined. Some of these may be new materials, while others may be artifacts produced during isolation by degradative conditions. It is also possible that certain components such as laminin may exist as a family of proteins including smaller variants.[94,107] Other components may have very specialized functions, e.g., as receptors for hormones or immune complexes or as factors that regulate the permeability of the basement membranes. Cell culture studies may help to explore this complexity as indicated by observations on some new disulfide-linked polypeptide chains synthesized in cultured tumor cells,[31] endothelial cells,[176] or epithelial cells.[177]

## IV. IMMUNOCHEMISTRY

Interest in the immunology of basement membranes was stimulated by the discovery of autoantibodies against basement membranes in certain human diseases,[167,171] including bullous pemphigoid and chronic glomerulonephritis. A variety of models have been used in experimental pathology to produce renal damage by injecting nephritogenic antigens or nephrotoxic antisera of heterologous origin. In most cases, the antigens used are complex, such as intact basement membranes or extracts of the glomeruli. These studies indicate that basement membranes contain strongly immunogenic components capable of eliciting autoimmune responses even when administered in insoluble form.[167,178] Related studies are now being undertaken with purified basement membrane proteins. Antibodies to Type IV collagen and laminin have been recently used to evaluate their immunopathological potency. Rabbit antisera against Type IV collagen or laminin when injected into mice were found to localize to renal and alveolar basement membranes and produced lesions resembling those in the Goodpasture syndrome,[179] and in addition, they caused a premature termination of pregnancy.[180]

The immunochemical characterization of Type IV collagen was carried out with pepsin-solubilized material from bovine lens capsule[181,182] and human placenta, as well as with acid-extracted material from the mouse EHS tumor.[183] Rabbit antisera reacted with both triple-helical molecules and their unfolded constituent chains, while guinea pig antibodies were directed against helical antigenic determinants. Antigenic determinants could be identified on the noncollagenous segments of the molecule such as shown in Figure 2 and on its triple-helical portion. Here, the N terminal portion (pepsin fragment P3) appeared to be a stronger antigen than the C terminal fragments P1 and P2. The antigenic structure of the latter has been characterized by radioimmunoassays.[184] Both fragments showed only weak cross-reaction (see Figure 13), indicating that $\alpha 1$(IV)- and $\alpha 2$(IV)-chains are structurally different. Mouse and human Type IV collagens have both unique and shared antigenic determinants which were localized to different cyanogen bromide peptides of the P1 fragment.

7-S collagens were also found to be strong immunogens.[17] The antisera reacted with the polypeptide chains obtained by reduction under denaturing conditions and showed restricted interspecies cross-reactions. 7-S collagens obtained from different organs of the same animal, however, were found to be identical antigens. A more extensive immunochemical characterization demonstrated several antigenic determinants either unique to the long or short form of the protein or shared by both.[185] The data also indicated several different polypeptide chains in 7-S collagen.

Antisera have also been prepared against Type V collagen[79–81] and the intima

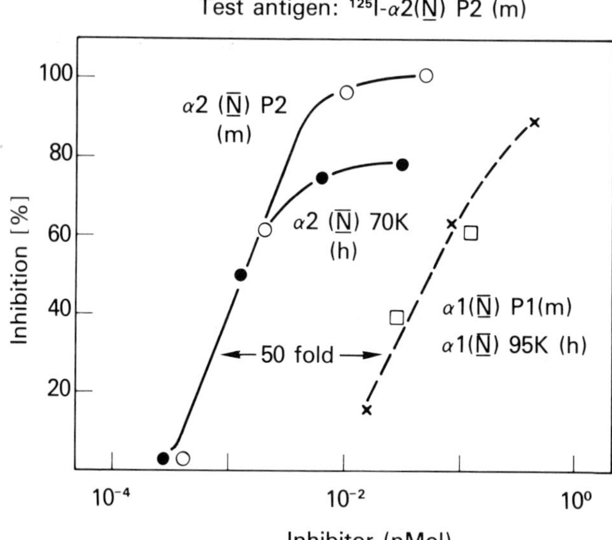


FIGURE 13. Comparison by radioimmunoassay of pepsin fragments from mouse (m) and human (h) Type IV collagen α1(IV)- and α2(IV)-chains. Rabbit antibodies to mouse Type IV collagen were preincubated with various inhibitors (P1, P2, 75K, 95K) prior to the incubation with labeled test antigen (indicated at the top). Note the almost identical reaction of pepsin fragments from mouse (○---○) and human (●---●) α2(IV)-chain, while similar fragments of α1(IV)-chain (×, ○) were about 50-fold less active. (Modified from Risteli, J., Schuppan, D., Glanville, R. W., and Timpl, R., *Biochem. J.*, 191, 517, 1980. With permission.)

collagen,[90] but these have not yet been extensively characterized. Rabbit antibodies against Type V collagen recognize helical antigenic determinants[79,186] which is a unique feature for rabbit antibodies against collagenous proteins.[187] Evidence exists, however, that the α1(V)-chain is also immunogenic.[188]

Laminin was found to be even a better immunogen than the basement membrane collagens.[93,101,102] The major antigenic structures of laminin require an intact conformation and were localized to fragment P1 derived from the center of the molecule. Other antigenic determinants were identified on smaller domains of the molecule such as pepsin fragment P2[107] and on the reduced and alkylated chain constituents. An immune response could also be elicited against the 400,000 and 200,000 dalton chain constituents isolated by electrophoresis.[95,100] At present the data are in conflict, showing distinct cross-reaction or lack of cross-reaction between both components. In spite of the considerable use of antisera against fibronectin, only little information is available on its immunochemical properties. The data so far indicate that the antigenic structure survives reduction and limited proteolysis and shows high interspecies cross-reactions.[115,189,190]

Several studies showed that the various basement membrane proteins are unique antigens which usually do not share antigenic determinants. Thus, it was possible to prepare specific antibodies for Types IV and V collagens for use in immunohistology.[79,191] It was also a major question whether similar proteins obtained from diverse sources, such as plasma, tumors, or cell cultures, were antigenically different. Comparisons made by absorption or immunochemical assays have not shown that Type IV collagen,[191] 7-S collagen,[17] or laminin[99,101] whether obtained from tumors, endodermal cells, or from authentic basement membranes are different. Complete cross-reactions were also found for fibronectins obtained from a variety of sources.[112] The preparation of monoclonal antibodies against some of these proteins is now in progress,[192] and these may be more useful reagents for detecting subtle differences between those proteins.

Well characterized antibodies to basement membrane components have been purified by affinity chromatography[79,93,163,191] and used to localize these proteins in tissues and cultured cells. At the light microscope level, complete codistribution was found with few exceptions for laminin, Type IV collagen, and the basement membrane proteoglycan[101,102,163] and also demonstrating that they were absent from interstitial connective tissues. Type V collagens showed either codistribution[80,81] with the basement membrane proteins or more restricted staining patterns, but also stained areas which lack basement membranes.[80,188] By using immunoelectronmicroscopy, laminin was localized to the lamina rara of epidermal[102] and glomerular[125,193] basement membranes, while Type IV collagen was mainly found in the lamina densa. Other basement membranes, such as in kidney tubules or mesangium,[81,125] showed a codistribution of laminin and Type IV collagen, indicating a different molecular architecture in these structures. These antibodies were also used for detecting Type IV collagen and laminin during early embryonic development.[194-198]

Radioimmunoassays and enzyme immunoassays have been developed for the detection of Type IV collagen, 7-S collagen, laminin, and laminin fragments in the nanogram range.[17,89,101,183,199] These assays were applied in quantitative studies on cell culture products[98,104,199] and for the detection of circulating antigens.[89] The latter approach may allow basement membrane turnover to be monitored *in situ* under normal and pathological conditions.

## REFERENCES


1. **Hay, E. D.,** *Epithelium* in Histology, Weiss, L. and Greep, R. O., Eds., McGraw-Hill, 1977, 113.
2. **Krakower, C. A. and Greenspon, S. A.,** The isolation of basement membranes, in *Biology and Chemistry of Basement Membranes*, Kefalides, N. A., Ed., Academic Press, New York, 1978, 1.
3. **Kefalides, N. A.,** Isolation of collagen from basement membrane containing three identical α-chains, *Biochem. Biophys. Res. Commun.*, 45, 226, 1971.
4. **Kefalides, N. A., Alper, R., and Clark, C. C.,** Biochemistry and metabolism of basement membranes, *Int. Rev. Cytol.*, 61, 167, 1979.
5. **Hudson, B. G. and Spiro, R. G.,** Fractionation of glycoprotein components of the reduced alkylated renal glomerular basement membrane, *J. Biol. Chem.*, 247, 4239, 1972.
6. **Farquhar, M. G.,** Structure and function in glomerular capillaries, in *Biology and Chemistry of Basement Membranes,* Kefalides, N. A., Ed., Academic Press, New York, 1978, 43.
7. **Piez, K. A.,** Primary structure, in *Biochemistry of Collagen,* Ramachandran, G. N. and Reddi, A. H., Eds., Plenum Press, New York, 1976, 1.
8. **Gay, S. and Miller, E. J.,** *Collagen in the Physiology and Pathology of Connective Tissue,* Gustav-Fischer Verlag, Stuttgart, 1978.
9. **Bornstein, P. and Sage, H.,** Structurally distinct collagen types, *Annu. Rev. Biochem.*, 49, 957, 1980.
10. **Kefalides, N. A.,** Structure and biosynthesis of basement membranes, *Int. Rev. Connect. Tissue Res.*, 6, 63, 1973.
11. **Sato, T. and Spiro, R. G.,** Studies on the subunit composition of the renal glomerular basement membrane, *J. Biol. Chem.*, 251, 4062, 1976.
12. **Spiro, R. G.,** Nature of the glycoprotein components of basement membranes, *Ann. N.Y. Acad. Sci.*, 312, 106, 1978.

13. **Grant, M. E., Kefalides, N. A., and Prockop, D. J.,** The biosynthesis of basement membrane collagen in embryonic chick lens. I. Delay between the synthesis of polypeptide and the secretion of collagen by matrix-free cells, *J. Biol. Chem.*, 247, 3539, 1972.
14. **Grant, M. E., Kefalides, N. A., and Prockop, D. J.,** The biosynthesis of basement membrane collagen in embryonic chick lens. II. Synthesis of a precursor form by matrix-free cells and a time dependent conversion to α chains in intact lens, *J. Biol. Chem.*, 247, 3545, 1972.
15. **Burgeson, R. E., El Adli, F. A., Kaitila, I. I., and Hollister, D. W.,** Fetal membrane collagens: identification of two new collagen alpha chains, *Proc. Natl. Acad. Sci. U.S.A.*, 73, 2579, 1976.
16. **Chung, E., Rhodes, R. K., and Miller, E. J.,** Isolation of three collagenous components of probable basement membrane origin from several tissues, *Biochem. Biophys. Res. Commun.*, 71, 1167, 1976.
17. **Risteli, J., Bächinger, H. P., Engel, J., Furthmayr, H., and Timpl, R.,** 7-S collagen: characterization of an unusual basement membrane structure, *Eur. J. Biochem.*, 108, 239, 1980.
18. **Furuto, D. K. and Miller, E. J.,** Isolation of a unique collagenous fraction from limited pepsin digests of human placental tissue, *J. Biol. Chem.*, 255, 290, 1980.
19. **Leblond, C. P., Glegg, R. E., and Eidinger, D.,** Presence of carbohydrates with free 1,2-glycol groups in sites stained by the periodic acid-Schiff technique, *J. Histochem. Cytochem.*, 5, 445, 1957.
20. **Pierce, G. B.,** Epithelial basement membrane: origin, development and role in disease, in *Chemistry and Molecular Biology of the Intracellular Matrix,* Balazs, E. A., Ed., Academic Press, 1970, 471.
21. **Orkin, R. W., Gehron, P., McGoodwin, E. B., Martin, G. R., Valentine, T., and Swarm, R.,** A murine tumor producing a matrix of basement membrane, *J. Exp. Med.*, 145, 204, 1977.
22. **Spiro, R. G.,** The structure of the disaccharide unit of the renal glomerular basement membrane, *J. Biol. Chem.*, 242, 4813, 1967.
23. **Spiro, R. G.,** Studies on the renal glomerular basement membrane. Nature of the carbohydrate units and their attachment to the peptide portion, *J. Biol. Chem.*, 242, 1923, 1967.
24. **Levine, M. J. and Spiro, R. G.,** Isolation from glomerular basement membrane of a glycopeptide containing both asparagine-linked and hydroxylysine-linked carbohydrate units, *J. Biol. Chem.*, 254, 8121, 1979.
25. **Timpl, R., Martin, G. R., Bruckner, P., Wick, G., and Wiedemann, H.,** Nature of the collagenous protein in a tumor basement membrane, *Eur. J. Biochem.*, 84, 43, 1978.
26. **Schuppan, D., Timpl, R., and Glanville, R. W.,** Discontinuities in the triple helical sequence Gly-X-Y of basement membrane (type IV) collagen, *FEBS Lett.*, 115, 297, 1980.
27. **Rauter, A. and Glanville, R. W.,** Pepsin fragments of human placental type IV collagen showing interrupted triple-helical amino acid sequences, *Hoppe-Seyler's Z. Physiol. Chem.*, 362, 943, 1981.
28. **Traub, W. and Piez, K. A.,** The chemistry and structure of collagen. *Adv. Protein Chem.*, 25, 243, 1971.
29. **Crouch, E. and Bornstein, P.,** Characterization of a type IV procollagen synthesized by human amniotic fluid cells in culture, *J. Biol. Chem.*, 254, 4197, 1979.
30. **Tryggvason, K., Gehron Robey, P., and Martin, G. R.,** Biosynthesis of type IV procollagens, *Biochemistry,* 19, 1284, 1980.
31. **Alitalo, K., Vaheri, A., Krieg, T., and Timpl, R.,** Biosynthesis of two subunits of type IV procollagen and of other basement membrane proteins by a human tumor cell line, *Eur. J. Biochem.*, 109, 247, 1980.
32. **Foidart, J. M., Tryggvason, K., Gehron Robey, P., Liotta, L. A., and Martin, G. R.,** Biosynthesis of type IV and V (αA and αB) collagens by human placenta, *Coll. Rec. Res.*, 1, 137, 1981.
33. **Fessler, L. I. and Fessler, J. H.,** Characterization of basement membrane procollagen made by human endothelial cells, *J. Supramol. Struct.*, 4, 178, 1980.
34. **Crouch, E., Sage, H., and Bornstein, P.,** Structural basis for apparent heterogeneity of collagens in human basement membranes: Type IV procollagen contains two distinct chains, *Proc. Natl. Acad. Sci. U.S.A.*, 77, 745, 1980.
35. **Olsen, B. R., Alper, R., and Kefalides, N. A.,** Structural characterization of a soluble fraction from lens capsule basement membrane, *Eur. J. Biochem.*, 38, 220, 1973.
36. **Gay, S. and Miller, E. J.,** Characterization of lens capsule collagen: evidence for the presence of two unique chains in molecules derived from major basement membrane structures, *Arch. Biochem. Biophys.*, 198, 370, 1979.
37. **Gehron Robey, P. and Martin, G. R.,** Type IV collagen contains two distinct chains in separate molecules, *Coll. Rec. Res.*, 1, 27, 1981.
38. **Schwartz, D., Chin-Quee, T., and Veis, A.,** Characterization of bovine anterior lens capsule basement membrane collagen. I. Pepsin susceptibility, salt precipitation and thermal gelation: a property of non-collagen component integrity, *Eur. J. Biochem.*, 103, 21, 1980.

39. **Schwartz, D. and Veis, A.,** Characterization of bovine anterior-lens-capsule basement-membrane collagen. II. Segment-long-spacing precipitates: further evidence for large N-terminal and C-terminal extensions, *Eur. J. Biochem.,* 103, 29, 1980.
40. **Butkowski, R. J., Todd, P., Grantham, J. J., and Hudson, B. G.,** Rabbit tubular basement membrane. Isolation and analysis of polypeptides, *J. Biol. Chem.,* 254, 10503, 1979.
41. **Freytag, J. W., Dalrymple, P. N., Maguire, M. H., Strickland, D. K., Carraway, K. L., and Hudson, B. G.,** Glomerular basement membrane. Studies on its structure and interaction with platelets, *J. Biol. Chem.,* 253, 9069, 1978.
42. **Tanzer, M. L. and Kefalides, N. A.,** Collagen crosslinks: occurrence in basement membrane collagen, *Biochem. Biophys. Res. Commun.,* 51, 775, 1973.
43. **Daniels, J. R. and Chu, G. H.,** Basement membrane collagen of renal glomerulus, *J. Biol. Chem.,* 250, 3531, 1975.
44. **Dixit, S. N.,** Isolation and characterization of two α chain size collagenous polypeptide chains C and D from glomerular basement membrane, *FEBS Lett.,* 106, 379, 1979.
45. **Tryggvason, K. and Kivirikko, K. I.,** Heterogeneity of pepsin-solubilized human glomerular basement membrane collagen, *Nephron,* 21, 230, 1978.
46. **Dixit, S. N.,** Type IV collagens. Isolation and characterization of two structurally distinct collagen chains from bovine kidney cortices, *Eur. J. Biochem.,* 106, 563, 1980.
47. **Dixit, S. N. and Kang, A. H.,** Basement membrane collagens. Cyanogen bromide peptides of the D chain from porcine kidney, *Biochemistry,* 19, 2692, 1980.
48. **Dehm, P. and Kefalides, N. A.,** The collagenous component of lens basement membrane. The isolation and characterization of an α chain size collagenous peptide and its relationship to newly synthesized lens components, *J. Biol. Chem.,* 253, 6680, 1978.
49. **Dixit, S. N.,** Isolation and characterization of two collagenous components from anterior lens capsule, *FEBS Lett.,* 85, 153, 1978.
50. **Dixit, S. N. and Kang, A. H.,** Anterior lens capsule collagens: cyanogen bromide peptides of the C chain, *Biochemistry,* 18, 5686, 1979.
51. **Bailey, A. J., Sims, T. J., Duance, V. C., and Light, N. D.,** Partial characterization of a second basement membrane collagen in human placenta. Evidence for the existence of two type IV collagen molecules, *FEBS Lett.,* 99, 361, 1979.
52. **Glanville, R. W., Rauter, A., and Fietzek, P. P.,** Isolation and characterization of a native placental basement membrane collagen and its component α chains, *Eur. J. Biochem.,* 95, 383, 1979.
53. **Kresina, T. F. and Miller, E. J.,** Isolation and characterization of basement membrane collagen from human placental tissue. Evidence for the presence of two genetically distinct collagen chains, *Biochemistry,* 18, 3089, 1979.
54. **Sage, H., Woodbury, R. G., and Bornstein, P.,** Structural studies on human type IV collagen, *J. Biol. Chem.,* 254, 9893, 1979.
55. **Mayne, R. and Zettergren, J. G.,** Type IV collagen from chicken muscular tissue. Isolation and characterization of the pepsin-resistant fragments, *Biochemistry,* 19, 4065, 1980.
56. **Timpl, R., Bruckner, P., and Fietzek, P. P.,** Characterization of pepsin fragments of basement membrane collagen obtained from a mouse tumor, *Eur. J. Biochem.,* 95, 255, 1979.
57. **Uitto, V-J., Schwartz, D., and Veis, A.,** Degradation of basement membrane collagen by neutral proteases from human leucocytes, *Eur. J. Biochem.,* 105, 409, 1980.
58. **Mainardi, C. L., Dixit, S. N., and Kang, A. H.,** Degradation of type IV (basement membrane) collagen by a proteinase from human polymorphonuclear leucocyte granules, *J. Biol. Chem.,* 255, 5435, 1980.
59. **Timpl, R., Risteli, J., and Bächinger, H. P.,** Identification of a new basement membrane collagen by the aid of a large fragment resistant to bacterial collagenase, *FEBS Lett.,* 101, 265, 1979.
60. **Woolley, D. E., Glanville, R. W., Roberts, D. R., and Evanson, J. M.,** Purification, characterization and inhibition of human skin collagenase, *Biochem. J.,* 169, 265, 1978.
61. **Timpl, R., Martin, G. R., and Bruckner, P.,** Structure of basement membrane collagen obtained from a mouse tumor, *Front. Matrix Biol.,* 7, 130, 1979.
62. **Liotta, L. A., Abe, S., Gehron Robey, P., and Martin, G. R.,** Preferential digestion of basement membrane collagen by an enzyme derived from a metastatic murine tumor, *Proc. Natl. Acad. Sci. U.S.A.,* 76, 2268, 1979.
63. **Hudson, B. G. and Spiro, R. G.,** Studies on the native and reduced and alkylated renal glomerular basement membrane. Solubility, subunit size, and reaction with cyanogen bromide, *J. Biol. Chem.,* 247, 4229, 1972.
64. **Kefalides, N. A.,** Isolation and characterization of cyanogen bromide peptides from basement membrane collagen, *Biochem. Biophys. Res. Commun.,* 47, 1151, 1972.

65. **Dixit, S. N., Seyer, J., Kang, A. H.,** Unpublished, 1981.
66. **Reid, K. B. M.,** Complete amino acid sequences of the three collagen-like regions present in subcomponent Clq of the first component of human complement, *Biochem. J.,* 179, 367, 1979.
67. **Hung, C.-H., Butkowski, R. J., and Hudson, B. G.,** Intestinal basement membrane of *Ascaris suum.* Properties of the collagenous domain, *J. Biol. Chem.,* 255, 4964, 1980.
68. **Gelman, R. A., Blackwell, J. Kefalides, N. A., and Tomichek, E.,** Thermal stability of basement membrane collagen, *Biochim. Biophys. Acta,* 427, 492, 1976.
69. **Roven, N., Ripamonti, A., Brigi, A., Volpin, D., and Giro, M. G.,** X-ray diffraction study of bovine lens capsule collagen, *Biochim. Biophys. Acta,* 576, 404, 1979.
70. **Trelstad, R. L. and Lawley, K. R.,** Isolation and initial characterization of human basement membrane collagen, *Biochem. Biophys. Res. Commun.,* 76, 376, 1977.
71. **Schwartz, D. and Veis, A.,** Characterization of basement membrane collagen of bovine anterior lens capsule via segment-long-spacing crystallites and the specific cleavage of the collagen by pepsin, *FEBS Lett.,* 85, 326, 1978.
72. **Schwartz, D. and Veis, A.,** Structure of bovine anterior lens capsule basement membrane collagen molecules from electron microscopy, *Biopolymers,* 18, 2363, 1979.
73. **Kühn, K., Wiedemann, H., Timpl, R., Risteli, J., Dieringer, H., Voss, T., and Glanville, R. W.,** Macromolecular structure of basement membrane collagens. Identification of 7-S collagen as the major cross-linking domain of basement membrane collagen, *FEBS Lett.,* 125, 123, 1981.
74. **Bentz, H., Bächinger, H. P., Glanville, R., and Kühn, K.,** Physical evidence for the assembly of A and B chains of human placental collagen in a single triple helix, *Eur. J. Biochem.,* 92, 563, 1978.
75. **Hong, B.-S, Davison, P. F., and Cannon, D. J.,** Isolation and characterization of a distinct type of collagen from bovine fetal membranes and other tissues, *Biochemistry,* 18, 4278, 1979.
76. **von der Mark, H. and von der Mark, K.,** Isolation and characterization of collagen A and B chains from chick embryos, *FEBS Lett.,* 99, 101, 1979.
77. **Sage, H. and Bornstein, P.,** Characterization of a novel collagen chain in human placenta and its relation to AB collagen, *Biochemistry,* 18, 3815, 1979.
78. **Brown, R. A., Shuttleworth, A., and Weiss, J. B.,** Three new α-chains of collagen from a non-basement membrane source, *Biochem. Biophys. Res. Commun.,* 80, 866, 1978.
79. **Madri, J. A. and Furthmayr, H.,** Isolation and tissue localization of type $AB_2$ collagen from normal lung parenchyma, *Am. J. Pathol.,* 94, 323, 1979.
80. **Duance, V. C., Restall, D. J., Beard, H., Bourne, F. J., and Bailey, A. J.,** The location of three collagen types in skeletal muscle, *FEBS Lett.,* 79, 248, 1977.
81. **Roll, F. J., Madri, J. A., Alberts, J., and Furthmayr, H.,** Codistribution of collagen types IV and $AB_2$ in basement membranes and mesangium of the kidney, *J. Cell. Biol.,* 85, 597, 1980.
82. **Pöschl, A. and von der Mark, K.,** Synthesis of type V collagen by chick corneal fibroblasts in vivo and in vitro, *FEBS Lett.,* 115, 100, 1980.
83. **Jimenez, S. A., Yankowski, R., and Bashey, R. I.,** Identification of two new collagen α-chains in extracts of lathyritic chick embryo tendons, *Biochem. Biophys. Res. Commun.,* 81, 1298, 1978.
84. **Burgeson, R. E. and Hollister, D. W.,** Collagen heterogeneity in human cartilage: identification of several new collagen chains, *Biochem. Biophys. Res. Commun.,* 87, 1124, 1979.
85. **Brown, R. A. and Weiss, J. B.,** Type V collagen: possible shared identity of αA, αB and αC chains, *FEBS Lett.,* 106, 71, 1979.
86. **Rhodes, R. K. and Miller, E. J.,** Physiochemical characterization and molecular organization of the collagen A and B chains, *Biochemistry,* 17, 3442, 1978.
87. **Rhodes, R. K. and Miller, E. J.,** The isolation and characterization of the cyanogen bromide peptides from the B chain of human collagen, *J. Biol. Chem.,* 254, 12084, 1979.
88. **Kumamoto, C. A. and Fessler, J. H.,** Biosynthesis of AB procollagen, *Proc. Natl. Acad. Sci. U.S.A.,* 77, 6434, 1980.
89. **Risteli, J., Rohde, H., and Timpl, R.,** Sensitive radioimmunoassays for 7-S collagen and laminin. Application to serum and tissue studies of basement membranes, *Anal. Biochem.,* 113, 372, 1981.
90. **Jander, R., Rauterberg, J., Voss, B., and von Bassewitz, D. B.,** A cysteine-rich protein from bovine placenta. Isolation of its constituent polypeptide chains and some properties of the non-denatured protein, *Eur. J. Biochem.,* 114, 17, 1980.
91. **Rojkind, M., Giambrone, M.-A., and Biempica, L.,** Collagen types in normal and cirrhotic liver, *Gastroenterology,* 76, 710, 1979.
92. **Shibata, S., Nagasawa, T., and Miura, K.,** Nephritogenoside, the receptor glycoprotein for concanavalin A in rat glomerular basement membrane, *Biochim. Biophys. Acta,* 499, 392, 1977.
93. **Timpl, R., Rohde, H., Gehron Robey, P., Rennard, S. I., Foidart, J. M., and Martin, G. R.,** Laminin — a glycoprotein from basement membranes, *J. Biol. Chem.,* 254, 9933, 1979.

94. **Hogan, B. L. M., Cooper, A. R., and Kurkinen, M.,** Incorporation into Reichert's membrane of laminin-like extracellular proteins synthesized by parietal endoderm cells of the mouse embryo, *Dev. Biol.,* 80, 289, 1980.
95. **Sakashita, S. and Ruoslahti, E.,** Laminin-like glycoproteins in extracellular matrix of endodermal cells, *Arch. Biochem. Biophys.,* 205, 283, 1980.
96. **Howe, C. C. and Solter, D.,** Identification of non-collagenous basement membrane glycopeptides synthesized by mouse parietal entoderm and an entodermal cell line, *Dev. Biol.,* 77, 480, 1980.
97. **Strickland, S., Smith, K. K., and Marotti, K. R.,** Hormonal induction of differentiation in teratocarcinoma skin cells: generation of parietal endoderm by retinoic acid and dibutyryl cAMP, *Cell,* 21, 347, 1980.
98. **Leivo, I., Alitalo, K., Risteli, L., Vaheri, A., Timpl, R., and Wartiovaara, J.,** Basal lamina glycoproteins laminin and type IV collagen are assembled into a fine-fibered matrix in cultures of a teratocarcinoma-derived endodermae cell line, *Exp. Cell Res.,* 137, 15, 1982.
99. **Sakashita, S., Engvall, E., and Ruoslahti, E.,** Basement membrane glycoprotein laminin binds to heparin, *FEBS Lett.,* 116, 243, 1980.
100. **Chung, A. E., Jaffe, R., Freeman, I. L., Vergnes, J.-P., Bragiuski, J. E., and Carlin, B.,** Properties of a basement membrane related glycoprotein synthesized in culture by a mouse embryonal carcinoma-derived line, *Cell,* 16, 277, 1979.
101. **Rohde, H., Wick, G., and Timpl, R.,** Immunochemical characterization of the basement membrane glycoprotein laminin, *Eur. J. Biochem.,* 101, 195, 1979.
102. **Foidart, J. M., Bere, E. W., Yaar, M., Rennard, S. I., Gullino, M., Martin, G. R., and Katz, S. I.,** Distribution and immunoelectron microscopic localization of laminin, a non-collagenous basement membrane glycoprotein, *Lab. Invest.,* 42, 336, 1980.
103. **Timpl, R., Rohde, H., Ott-Ulbricht, U., Risteli, L., and Bächinger, H. P.,** Chemical characterization of laminin, a major glycoprotein of basement membranes, in *Glycoconjugates,* Schauer, R., Boer, P., Buddecke, E., Kramer, M. F., Vliegenthart, J. F. G., and Wiegandt, H., Eds., George Thieme, Stuttgart, 1979, 145.
104. **Alitalo, K., Kurkinen, M., Vaheri, A., Krieg, T., and Timpl, R.,** Extracellular matrix components synthesized by human amniotic epithelial cells in culture, *Cell,* 19, 1053, 1980.
105. **Engel, J., Odermatt, E., Engel, A., Madri, J. A., Furthmayr, H., Rohde, H., and Timpl, R.,** Shapes domain organization and flexibility of laminin and fibronectin; two multifunctional extracellular matrix proteins, *J. Molec. Biol.,* 150, 97, 1981.
106. **Ott, U., Odermatt, E., Engel, J., Furthmayr, H., and Timpl, R.,** Protease resistance and conformation of laminin, *Eur. J. Biochem.,* in press, 1982.
107. **Rohde, H., Bächinger, H. P., and Timpl, R.,** Characterization of pepsin fragments of laminin in a tumor basement membrane. Evidence for the existence of related proteins, *Hoppe-Seylers Z. Physiol. Chem.,* 361, 1651, 1980.
108. **Risteli, L. and Timpl, R.,** Isolation and characterization of pepsin fragments of laminin from human placental and renal basement membranes, *Biochem. J.,* 193, 749, 1981.
109. **Terranova, V. P., Rohrbach, D. H., and Martin, G. R.,** Role of laminin in the attachment of PAM 212 (epithelial) cells to basement membrane collagen, *Cell,* 22, 719, 1980.
110. **Vaheri, A. and Mosher, D. F.,** High molecular weight, cell surface associated glycoprotein (fibronectin) lost in malignant transformation, *Biochim. Biophys. Acta,* 516, 1, 1978.
111. **Vaheri, A., Ruoslahti, E., and Mosher, D. F., Eds.,** Fibroblast surface protein, *Ann. N.Y. Acad. Sci.,* 312, 1, 1978.
112. **Mosher, D. F.,** Fibronectin, *Prog. Hemostasis Thromb.,* 5, 111, 1980.
113. **Yamada, K. M. and Olden, K.,** Fibronectins-adhesive glycoproteins of cell surface and blood, *Nature,* 275, 179, 1978.
114. **Pearlstein, E., Gold, L. I., and Garcia-Pardo, A.,** Fibronectin: a review of its structure and biological activity, *Mol. Cell. Biochem.,* 29, 103, 1980.
115. **Hynes, R. O.,** Cell-surface proteins and malignant transformation, *Biochim. Biophys. Acta,* 458, 73, 1976.
116. **Ruoslahti, E., Engvall, E., and Hayman, E. G.,** Fibronectin: current concepts of its structure and function, *Coll. Rec. Res.,* 1, 95, 1981.
117. **Ruoslahti, E. and Vaheri, A.,** Novel human serum protein from fibroblast plasma membranes, *Nature,* 248, 789, 1974.
118. **Ruoslahti, E. and Vaheri, A.,** Interaction of soluble fibroblast surface antigen with fibrinogen and fibrin. Identity with cold-insoluble globulin of human plasma, *J. Exp. Med.,* 141, 497, 1975.
119. **Mosesson, M. W. and Umfleet, R. A.,** The cold-insoluble globulin of human plasma, *J. Biol. Chem.,* 245, 5728, 1970.

120. **Crouch, E., Balian, G., Holbrook, K., Duksin, D., and Bornstein, P.,** Amniotic fluid fibronectin. Characterization and synthesis by cells in culture, *J. Cell Biol.*, 78, 701, 1978.
121. **Yamada, K. M. and Kennedy, D.,** Fibroblast cellular and plasma fibronectins are similar but not identical, *J. Cell Biol.*, 80, 492, 1979.
122. **Pena, S. D. J., Mills, G., Hughes, C., and Aplin, J. D.,** Polypeptide heterogeneity of hamster and calf fibronectins, *Biochem. J.*, 189, 337, 1980.
123. **Wartiovaara, J., Leivo, I., and Vaheri, A.,** Expression of the cell surface-associated glycoprotein, fibronectin, in the early mouse embryo, *Dev. Biol.*, 69, 247, 1979.
124. **Stenman, S. and Vaheri, A.,** Distribution of a major connective tissue protein, fibronectin, in normal human tissue, *J. Exp. Med.*, 147, 1054, 1978.
125. **Madri, J. A., Roll, J., Furthmayr, H., and Foidart, J. M.,** Ultrastructural localization of fibronectin and laminin in the basement membranes of the murine kidney, *J. Cell Biol.*, 86, 682, 1980.
126. **Cortoy, P. J., Kanwar, Y. S., Hynes, R. O., and Farquhar, M. G.,** Fibronectin localization in the rat glomerulus, *J. Cell Biol.*, 87, 691, 1980.
127. **Hedman, K., Vaheri, A., and Waartiovaara, J.,** External fibronectin of cultured human fibroblasts is predominantly a matrix protein, *J. Cell Biol.*, 76, 748, 1978.
127a. **Furcht, H., Smith, D., Wendelschafer-Crabb, G., Mosher, D. F., and Foidart, J. M.,** Fibronectin presence in native collagen fibrils of human fibroblasts: immunoperoxidase and immunoferritin localization, *J. Histochem. Cytochem.*, 28, 1319, 1980.
128. **Vuento, M., Vartio, T., Saraste, M., von Bonsdorff, C. H., and Vaheri, A.,** Spontaneous and polyamine-induced formation of filamentous polymers from soluble fibronectin, *Eur. J. Biochem.*, 105, 33, 1980.
129. **Mosher, D. F., Schad, P. E., and Kleinman, H. K.,** Cross-linking of fibronectin to collagen by blood coagulation factor $XIII_a$, *J. Clin. Invest.*, 64, 781, 1979.
129a. **Yamada, K. M., Schlesinger, D. H., Kennedy, D. W., and Pastan, I.,** Characterization of a major fibroblast cell surface glycoprotein, *Biochemistry*, 16, 5552, 1977.
130. **Colonna, G., Alexander, S. S., Yamada, K. M., Pastan, I., and Edelhoch, H.,** The stability of cell surface protein to surfactants and denaturants, *J. Biol. Chem.*, 253, 7787, 1978.
131. **Mosesson, M. W., Chen, A. B., and Huseby, R. M.,** The cold-insoluble globulin of human plasma: studies of its essential structural features, *Biochim. Biophys. Acta*, 386, 509, 1975.
132. **Furie, M. B. and Rifkin, D. B.,** Proteolytically derived fragments of human plasma fibronectin and their localization within the intact molecule, *J. Biol. Chem.*, 255, 3134, 1980.
133. **Iwanaga, S., Suzuki, K., and Hashimoto, S.,** Bovine plasma cold-insoluble globulin: gross structure and function, *Ann. N.Y. Acad. Sci.*, 312, 56, 1978.
134. **Takasaki, S., Yamashita, K., Suzuki, K., Iwanaga, S., and Kobata, A.,** The sugar chains of cold-insoluble globulin, *J. Biol. Chem.*, 254, 8548, 1979.
135. **Fukuda, M. and Hakamori, S.,** Carbohydrate structure of galactoprotein a, a major transformation-sensitive glycoprotein released from hamster embryo fibroblasts, *J. Biol. Chem.*, 254, 5451, 1979.
136. **Alexander, S. S., Colonna, G., and Edelhoch, H.,** The structure and stability of human plasma cold-insoluble globulin, *J. Biol. Chem.*, 254, 1501, 1979.
137. **Mosher, D. F.,** Cross-linking of cold-insoluble globulin by fibrin-stabilizing factor, *J. Biol. Chem.*, 250, 6614, 1975.
138. **Jilek, F. and Hörmann, H.,** Cold-insoluble globulin: plasminolysis of cold-insoluble globulin, *Hoppe-Seylers Z. Physiol. Chem.*, 358, 1165, 1977.
139. **Wagner, D. D. and Hynes, R. O.,** Domain structure of fibronectin and its relation to function, *J. Biol. Chem.*, 254, 6746, 1979.
140. **Fukuda, M. and Hakamori, S.,** Proteolytic and chemical fragmentation of galactoprotein a, a major transformation-sensitive glycoprotein released from hamster embryo fibroblasts, *J. Biol. Chem.*, 254, 5442, 1979.
141. **Hahn, L. H. E. and Yamada, K. M.,** Isolation and biological characterization of active fragments of the adhesive glycoprotein fibronectin, *Cell*, 18, 1043, 1979.
142. **Ruoslahti, E., Hayman, E. G., Kuusela, P., Shively, J. E., and Engvall, E.,** Isolation of a tryptic fragment containing the collagen-binding site of plasma fibronectin, *J. Biol. Chem.*, 254, 6054, 1979.
143. **Sekiguchi, K. and Hakamori, S. I.,** Functional domain structure of fibronectin, *Proc. Natl. Acad. Sci. U.S.A.*, 77, 2661, 1980.
144. **Wagner, D. D. and Hynes, R. O.,** Topological arrangement of the major structural features of fibronectin, *J. Biol. Chem.*, 255, 4304, 1980.
145. **McDonald, J. A. and Kelley, D. G.,** Degradation of fibronectin by human leukocyte elastase. Release of biologically active fragments, *J. Biol. Chem.*, 255, 8848, 1980.
146. **Yamada, K. M., Kennedy, D. W., Kimata, K., and Pratt, R. M.,** Characterization of fibronectin interactions with glycosaminoglycans and identification of active proteolytic fragments, *J. Biol. Chem.*, 255, 6055, 1980.

147. **Balian, G., Click, E. M., and Bornstein, P.,** Location of a collagen-binding domain in fibronectin, *J. Biol. Chem.*, 255, 3234, 1980.
148. **Furie, M. B., Frey, A. B., and Rifkin, D. B.,** Location of a gelatin-binding region of human plasma fibronectin, *J. Biol. Chem.*, 255, 4391, 1980.
149. **Mosher, D. F., Schad, P. E., and Vann, J. M.,** Cross-linking of collagen and fibronectin by factor $XIII_a$: localization of participating glutaminyl residues to a tryptic fragment of fibronectin, *J. Biol. Chem.*, 255, 1181, 1980.
150. **Engvall, E., Ruoslahti, E., and Miller, E. J.,** Affinity of fibronectin to collagens of different genetic types and to fibrinogen, *J. Exp. Med.*, 147, 1584, 1978.
151. **Dessau, W., Adelmann, B. C., Timpl, R., and Martin, G. R.,** Identification of the sites in collagen α-chains that bind serum anti-gelatin factor (cold-insoluble globulin), *Biochem. J.*, 169, 55, 1978.
152. **Jilek, F. and Hörmann, H.,** Cold-insoluble globulin (fibronectin): affinity to soluble collagen of various types, *Hoppe-Seylers Z. Physiol. Chem.*, 359, 247, 1978.
153. **Kleinman, H. K., McGoodwin, E. B., Martin, G. R., Klebe, R. J., Fietzek, P. P., and Woolley, D. E.,** Localization of the binding site for cell attachment in the α1(I) chain of collagen, *J. Biol. Chem.*, 253, 5642, 1978.
154. **Ruoslahti, E. and Hayman, E. G.,** Two active sites with different characteristics in fibronectin, *FEBS Lett.*, 97, 221, 1979.
155. **Molnar, J., Gelder, F. B., Lai, M. Z., Siefring, G. E., Credo, R. B., and Lorand, L.,** Purification of opsonically active human and rat cold-insoluble globulin (plasma fibronectin), *Biochemistry*, 18, 3909, 1979.
156. **Kleinman, H. K., Martin, G. R., and Fishman, P. H.,** Ganglioside inhibition of fibronectin-mediated cell adhesion to collagen, *Proc. Natl. Acad. Sci. U.S.A.*, 76, 3367, 1979.
157. **Hay, E. D. and Meier, S.,** Glycosaminoglycan synthesis by embryonic inductors: neural tube, notochord and lens, *J. Cell Biol.*, 62, 889, 1974.
158. **Conn, R. H., Bannerjee, S. D., and Bernfield, M. R.,** Basal lamina of embryonic salivary epithelia, *J. Cell Biol.*, 73, 464, 1977.
159. **Trelstad, R. L., Hayashi, K., and Toole, B. P.,** Epithelial collagens and glycosaminoglycans in the embryonic cornea. Macromolecular order and morphogenesis in the basement membrane, *J. Cell Biol.*, 62, 815, 1974.
159a. **Kanwar, Y. S. and Farquhar, M. G.,** Anionic sites in the glomerular basement membrane. In vivo and in vitro localization to the laminae rarae by cationic probes, *J. Cell Biol.*, 81, 137, 1979.
160. **Kanwar, Y. S. and Farquhar, M. G.,** Presence of heparan sulfate in the glomerular basement membrane, *Proc. Natl. Acad. Sci. U.S.A.*, 76, 1303, 1979.
161. **Kanwar, Y. S. and Farquhar, M. G.,** Isolation of glycosaminoglycans (heparan sulfate) from glomerular basement membranes, *Proc. Natl. Acad. Sci. U.S.A.*, 76, 4493, 1979.
162. **Cohen, M. P.,** Glycosaminoglycans are integral constituents of renal glomerular basement membrane, *Biochem. Biophys. Res. Commun.*, 92, 343, 1980.
163. **Hassell, J. R., Gehron Robey, P., Barrach, H. J., Wilczek, J., Rennard, S. I., and Martin, G. R.,** A basement membrane proteoglycan isolated from the EHS sarcoma, *Proc. Natl. Acad. Sci. U.S.A.*, 77, 4494, 1980.
164. **Hahn, E., Wick, G., Pencev, D., and Timpl, R.,** Distribution of basement membrane proteins in normal and fibrotic human liver: collagen type IV, laminin and fibronectin, *Gut*, 21, 63, 1980.
165. **Johnson, L. D. and Warfel, J.,** Isolation and characterization of an epithelial basement membrane glycoprotein from murine kidney and further characterization of an epithelial basement membrane glycoprotein secreted by murine teratocarcinoma cells in vitro, *Biochim. Biophys. Acta*, 455, 538, 1976.
166. **Johnson, L. D. and Starcher, B. C.,** Epithelial basement membranes: the isolation and identification of a soluble component, *Biochim. Biophys. Acta*, 290, 158, 1972.
167. **Wilson, C. B. and Dixon, F. J.,** Immunopathology and glommerular nephritis, *Annu. Rev. Med.*, 25, 83, 1974.
168. **Marquardt, H., Wilson, C. B., and Dixon, F. J.,** Human glomerular basement membrane. Selective solubilization with chaotropes and chemical and immunologic characterization of its components, *Biochemistry*, 12, 3260, 1973.
169. **Mahieu, P. M., Lambert, P. H., and Maghuin-Rogister, G. R.,** Primary structure of a small glycopeptide isolated from human glomerular basement membrane and carrying a major antigenic site, *Eur. J. Biochem.*, 40, 599, 1973.
170. **Wick, G. and Timpl, R.,** Study on the nature of the Goodpasture antigen using a basement membrane producing mouse tumor, *Clin. Exp. Immunol.*, 39, 733, 1980.
171. **Beutner, E. H., Chorzelski, T. P., and Jordon, R. E.,** *Autosensitization in Pemphigus and Bullous Pemphigoid*, Charles C Thomas, Springfield, Ill., 1971.

172. **Diaz, L. A., Calvanico, N. J., Tomasi, T. B., and Jordon, R. E.,** Bullous pemphigoid antigen: isolation from normal human skin, *J. Immunol.*, 118, 455, 1977.
173. **Bardos, P., Muh, J. P., Luthier, B., Devulder, B., and Tacquet, B.,** Immunochemical study of some glycoproteins of the rat glomerular basement membrane, *Comp. Biochem. Physiol.*, 53, 49, 1976.
174. **Kefalides, N. A.,** The chemistry of antigenic components isolated from glomerular basement membrane, *Connect. Tissue Res.*, 1, 3, 1972.
175. **Huang, F. and Kalant, N.,** Isolation and characterization of antigenic components of rat glomerular basement membrane, *Can. J. Biochem.*, 46, 1523, 1968.
176. **Sage, H., Pritzl, P., and Bornstein, P.,** A unique pepsin sensitive collagen synthesized by aortic endothelial cells in culture, *Biochemistry*, 19, 5747, 1980.
177. **Quaromi, A. and Trelstad, R. L.,** Biochemical characterization of collagens synthesized by intestinal epithelial cell cultures, *J. Biol. Chem.*, 255, 8350, 1980.
178. **Merrill, J. P.,** Glomerulonephritis, *N. Engl. J. Med.*, 290, 257, 1974.
179. **Wick, G., Müller, P.-U., Timpl, R., and Martin, G. R.,** Studies on the immunology of basement membrane collagen using antibody to a tumor basement membrane, *Front. Matrix Biol.*, 7, 120, 1979.
180. **Foidart, J. M., Yaar, M., and Brown, K. S.,** Unpublished, 1981.
181. **Gunson, D. E. and Kefalides, N. A.,** The use of the radioimmunoassay in the characterization of antibodies to basement membrane collagen, *Immunology*, 31, 563, 1976.
182. **Gunson, D. E., Arbogast, B., and Kefalides, N. A.,** Rat parietal yolk sac basement membrane. An investigation of the antigenic determinants using a radioimmunoassay, *Immunology*, 31, 577, 1976.
183. **Timpl, R., Glanville, R. W., Wick, G., and Martin, G. R.,** Immunochemical study on basement membrane (type IV) collagens, *Immunology*, 38, 109, 1979.
184. **Risteli, J., Schuppan, D., Glanville, R. W., and Timpl, R.,** Immunochemical distinction between two different chains of type IV collagen, *Biochem. J.*, 191, 517, 1980.
185. **Risteli, J., Wick, G., and Timpl, R.,** Immunological characterization of the 7-S domain of type N collagens, *Coll. Rec. Res.*, 1, 419, 1982.
186. **Risteli, L., Bentz, H., and Timpl, R.,** unpublished, 1981.
187. **Furthmayr, H. and Timpl, R.,** Immunochemistry of collagens and procollagens, *Int. Rev. Connect. Tissue Res.*, 7, 61, 1976.
188. **Gay, S., Rhodes, R. K., Gay, R. E., and Miller, E. J.,** Collagen molecules comprised of α1(V)-chains (B-chains): an apparent localization in the exocytoskeleton, *Coll. Rec. Res.*, 1, 53, 1981.
189. **Kuusela, P., Ruoslahti, E., Engvall, E., and Vaheri, A.,** Immunological interspecies cross-reactions of fibroblast surface antigen (fibronectin), *Immunochemistry*, 13, 639, 1976.
190. **Ruoslahti, E. and Engvall, E.,** Immunochemical and collagen-binding properties of fibronectin, *Ann. N.Y. Acad. Sci.*, 312, 178, 1978.
191. **Wick, G., Glanville, R. W., and Timpl, R.,** Characterization of antibodies to basement membrane (type IV) collagen in immunohistological studies, *Immunobiology*, 156, 372, 1979.
192. **Zardi, L., Carmemolla, B., Siri, A., Santi, L., and Accolla, R. S.,** Somatic cell hybrids producing antibodies specific to human fibronectin, *Int. J. Cancer*, 25, 325, 1980.
193. **Courtoy, P. J., Kanwar, Y. S., Timpl, R., Hynes, R. O., and Farquhar, M. G.,** Comparative distribution of fibronectin, type IV collagen and laminin in the rat glomerulus, *J. Cell Biol.*, 87 (Abstr.), E943, 1980.
194. **Adamson, E. D. and Ayers, S. E.,** The localization and synthesis of some collagen types in developing mouse embryos, *Cell*, 16, 953, 1979.
195. **Sherman, M. I., Gay, R., Gay, S., and Miller, E. J.,** Association of collagen with preimplantation and peri-implantation mouse embryos, *Dev. Biol.*, 74, 470, 1980.
196. **Ekblom, P., Alitalo, K., Vaheri, A., Timpl, R., and Saxen L.,** Induction of a basement membrane glycoprotein in embryonic kidney: possible role of laminin in morphogenesis, *Proc. Natl. Acad. Sci. U.S.A.*, 77, 485, 1980.
197. **Leivo, I., Vaheri, A., Timpl, R., and Wartiovaara, J.,** Appearance and distribution of collagens and laminin in the early mouse embryo, *Dev. Biol.*, 76, 100, 1980.
198. **Ekblom, P., Lehtonen, E., Saxen, L., and Timpl, R.,** Shift in collagen type as an early response to induction of the metanephric mesenchyme, *J. Cell Biol.*, 89, 276, 1981.
199. **Rennard, S. I., Berg, R., Martin, G. R., Foidart, J. M., and Gehron Robey, P.,** Enzyme-linked immunoassay (ELISA) for connective tissue components, *Anal. Biochem.*, 104, 205, 1980.
200. **Glanville, R. W. and Schuppan, D.,** Unpublished, 1981.
201. **Schuppan, D., Glanville, R. W., and Timpl, R.,** Unpublished, 1981.
202. **Timpl, R., Wiedemann, H., van Deeden, V., Furthmayr, H., and Kühn, K.,** A network model for the organization of type N collagen molecules in basement membranes, *Eur. J. Biochem.*, 120, 203, 1981.
203. **Risteli, J., Odermatt, E., Furthmayr, H., Engel, J., and Timpl, R.,** Unpublished, 1981.

Chapter 6

# COLLAGENOUS MATRICES AS DETERMINANTS OF CELL FUNCTION

**H. K. Kleinman, D. H. Rohrbach, V. P. Terranova, H. H. Varner, A. T. Hewitt, G. R. Grotendorst, C. M. Wilkes, G. R. Martin, H. Seppä, and E. Schiffmann**

## TABLE OF CONTENTS

# I. INTRODUCTION

Collagen, a significant component of almost all tissues, has long been recognized as the major structural element in the body.[1-4] Although it has been described as a "biological rope" which gives tissues their tensile strength, it is now clear that collagen either alone or with other matrix molecules also has important influences on cell growth and function. Many early studies suggested the importance of the extracellular matrix in the differentiation and migration of cells.[5-12] Recently, the molecular basis underlying the interaction of cells with collagen has been better defined, and new experimental systems are available for evaluating the nonstructural functions of collagens.[13-16]

Collagenous matrices appear differently at both the histological and ultrastructural levels in various tissues as, for example, in tendons, cartilage, and in basement membranes (see Figure 1). The collagen fibers in skin and tendon are large and have a very distinct band pattern (see Figure 1A). The fibers in cartilage around chondrocytes are finer and their appearance is partly obscured by the platelike deposits formed from the cartilage proteoglycan as a result of tissue fixation and staining (see Figure 1B). The basement membranes contain both an electron lucid and a dense zone. The basement membrane collagen is localized in the dense zone as a fine, nonfibrillar deposit.[17] When basement membranes are stained with ruthenium red, which localizes to anionic groups, the basement membrane proteoglycan with heparan sulfate chains is localized along the collagenous zone [18-20] (see Figure 1C).

A significant factor underlying the differences in the matrices is that they are largely composed of genetically and chemically distinct collagen types.[1-4] Five collagens have been defined, while others undoubtedly exist. Type I collagen is the most abundant and is found in many sites, including skin, tendon, bone, and the supportive tissues of most organs. Type II collagen is found in cartilage and in the vitreous humor of the eye and is produced by chondrocytes and by cells in the neural retina. Type III collagen is prominent in blood vessels, spleen, placenta, skin, and many of the tissues which contain Type I collagen except normal bone. Type III collagen is produced by fibroblasts and by smooth muscle cells. Type IV collagen is found specifically in basement membranes and is produced by epithelial and endothelial cells. Type V ($\alpha A \alpha B$) collagen is present in blood vessels, placenta, and several other tissues. The histological distribution of this collagen is not yet well established. Thus, some collagens, such as Types I, III, and V, are produced by a variety of cells and are found in various tissues, while others, such as Types II and IV, are produced by a limited number of cells and are tissue-specific. These collagens are associated with other matrix molecules, such as proteoglycans,[21,22] and attachment proteins, including fibronectin,[13,14,23-26] chondronectin,[27] and laminin.[28] The interactions of these molecules generate unique matrices.

# II. ROLE OF COLLAGEN IN CELLULAR METABOLISM AND DIFFERENTIATION

## A. Differentiation

It has been known for some time that many cells show improved growth and differentiation in culture on collagen substrates compared to that seen on glass or plastic substrates (Table 1).[6] Examination of the cell-collagen interaction has been limited perhaps because many cells grow adequately on the plastic substrate of tissue culture dishes. The nature of the interaction between the cell surface and the plastic substrate is not well defined. However, recent studies suggest that the plastic surface

**Table 1**
**INFLUENCE OF MATRIX PROTEINS ON DIFFERENTIATION**

| Cell or tissue | Matrix molecule required | Differentiated product | Ref. |
|---|---|---|---|
| Effect of collagen | | | |
| Chick myoblasts | Dried collagen film | Multinucleated myotubes | 7 |
| Corneal fibroblasts | Dried collagen film | Stroma production, EGF responsiveness | 9,12 |
| Guinea pig epidermal cells | Dried collagen Type IV film | Squamous epithelium formation | 29 |
| Chick somites | Soluble Type II collagen | Cartilage formation | 30 |
| Effect of collagenous matrices | | | |
| Subcutaneous in vivo implant | Dried decalcified bone powder | Bone formation | 31 |
| Muscle cells | Dried decalcified bone powder | Cartilage formation | 32 |
| Neural crest mesenchyme | Matrix of retinal pigmented epithelium | Cartilage formation | 33 |
| Corneal fibroblasts | Freeze-thawed lens | Stroma production | 9 |
| Effect of collagenase | | | |
| Chick skin tracts | Collagen | Collagenase blocks feather morphogenesis | 34 |
| Salivary rudiment | Collagen | Collagenase blocks branching salivary glands | 35 |
| Uteric rudiment | Collagen | Collagenase blocks normal uteric bud morphology | 36 |
| Effect of proline analogue-*L*-azedidine-)-decarboxylic acid | | | |
| Lung epithelium | Collagen | Analogue blocks epithelial branching | 37 |
| Tooth mesenchyme | Collagen | Analogue blocks ameloblast differentiation | 37 |
| Effect of pure fibronectin | | | |
| Chick myoblasts | Soluble exogenous fibronectin | Delays myogenesis | 38 |
| Neural crest cells | Soluble exogenous fibronectin | Blocks pigmentation and enhances adrenergic nerve cell differentiation | 39 |
| Chick chondrocytes | Soluble exogenous fibronectin | Blocks chondrogenesis | 40 |

has a high affinity for glycoproteins such as fibronectin and that the cells attach to these adsorbed proteins.[41–43] In this way, the plastic substrate is a substitute for collagen and for other matrix components which form the physiological substrate for these cells. It is possible that the types of cells that grow in culture as well as their phenotype are dependent on the substrate on which they are cultured. Evidence for this effect of collagen on cells is presented below.

*1. Effect of Purified Collagens*

Myoblasts adhere to plastic substrates, but do not form myotubes. However, when the dishes are coated with as little as 2 μg of collagen per 900 $mm^2$ dish, myoblasts will form multinucleated myotubes.[7] Collagen Types I to IV are equally active both in the native and denatured forms.[44] The active region of the Type I collagen molecule has been localized to α1(I)-CB7, a peptide produced from the α1(I)-chain by cyanogen bromide digestion.[45] This peptide also contains the fibro-

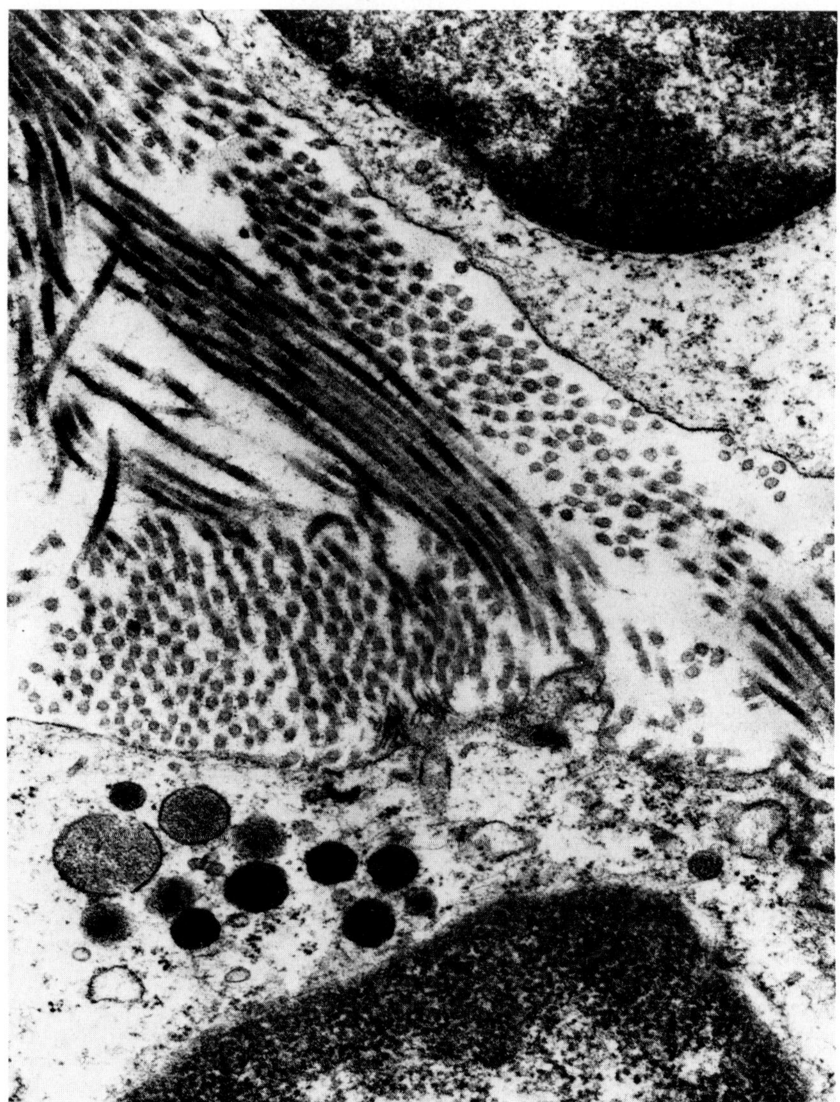

FIGURE 1. Electron micrographs of certain extracellular matrices. (A.) Lymph node cells and extracellular matrix with prominent Type I collagen fibrils. (Prepared by A. Anderson, University of Pennyslvania, University Park, Pa.)

nectin binding site.[23,24] Myoblasts, but not myotubes, contain significant amounts of fibronectin on their cell surface, suggesting that it could mediate their interaction with collagen.[46,47] Since added fibronectin delays or even prevents the fusion of myoblasts, the loss of cell surface fibronectin must be an obligatory step in their differentiation.[38]

The addition of soluble gelatin to cultured hog thyroid cells has been found to promote the reorganization of these cells into folliclelike structures.[48] Collagen can thus replace thyrotropin in promoting these cells to differentiate. In the presence of low concentrations of the hormone, the addition of collagen increases the response. Collagen may be exerting its effect by interacting directly with the cell, with

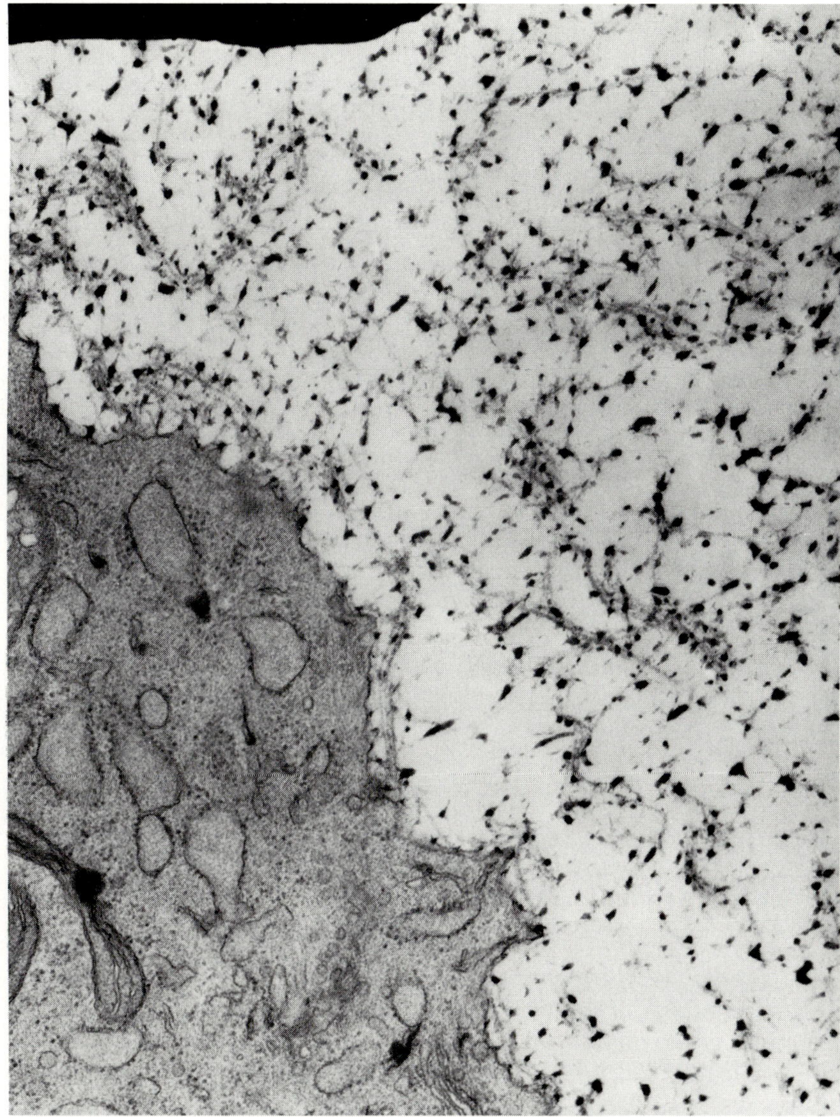

FIGURE 1B. Chondrocyte (lower left) and its etracellular matrix stained with ruthenium red which causes the proteoglycans to appear as small dark granules. The collagen appears as strands. (Prepared by G. Hascall, National Institute of Dental Research.)

fibronectin or with other serum components present in the culture fluid which then bind to the cell surface.

Isolated basal epidermal cells from skin undergo terminal differentiation when placed in suspension culture.[49] The changes include an increase in membrane permeability, the disulfide cross-linking of keratin, the formation of a cornified envelope, and the loss of the cell nucleus. Similar changes are seen in the cells in the stratum corneum. The critical step in the differentiation of these cells apparently is their separation from the basement membrane. When cultured on a feeder layer of cells or on collagen, the cells attach and their terminal differentiation is delayed.[50,51] It has recently been found that these cells have a decided preference for a substrate

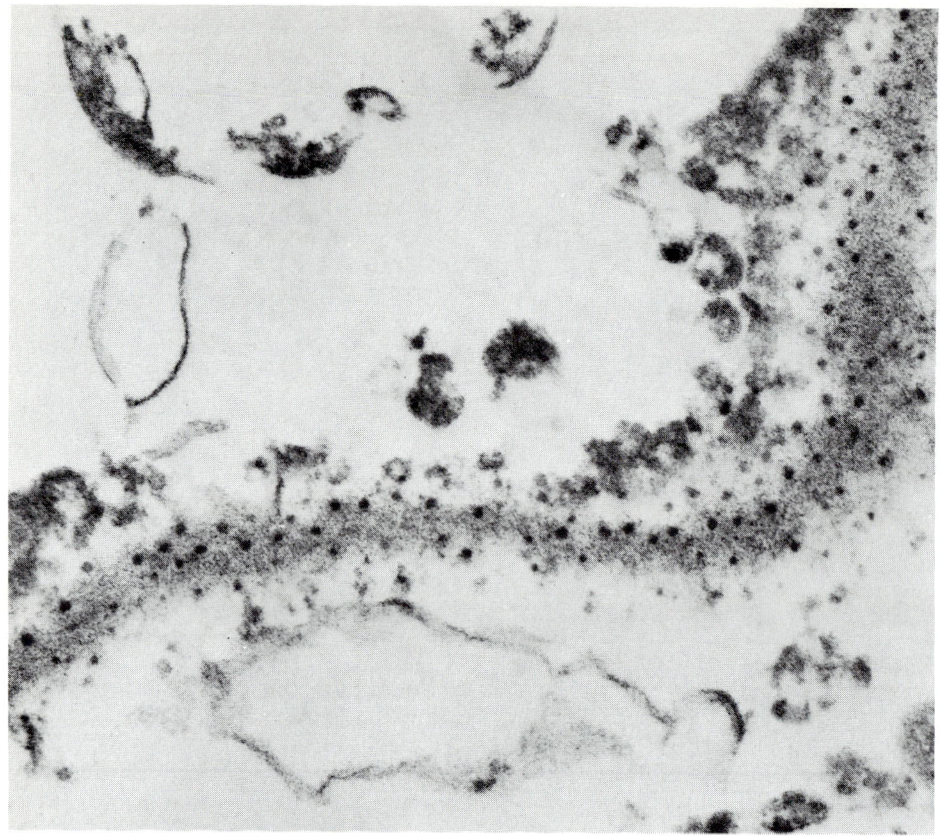

FIGURE 1C. Basement membranes of pulmonary alveolus stained with ruthenium red display prominent heparan sulfate proteoglycan granules in the edge of the lamina densa. (From Vaccaro, C. A. and Brody, J. B., *Am. Rev. Respir. Dis.*, 120, 901, 1979. With permission.)

of Type IV collagen.[29] They attach to the Type IV collagen using laminin[52] and proliferate and differentiate to form a multilayered squamous epithelium.[29] Thus, purified basement membrane collagen can support these cells in a way that allows both proliferation and differentiation.

Lastly, corneal epithelial cells show unique responses to collagen. Purified collagen substrates or freeze-thawed lens promote the differentiation of these cells and allow them to produce stroma.[9] The shape and growth of these cells is also under the influence of a collagen substrate. When maintained on plastic, they are flattened and do not respond to epidermal growth factor.[12] However, when maintained on collagen, they extend and assume a columnar shape and can respond to epidermal growth factor. Thus, collagen matrices not only control cell adhesion and differentiation, but may also be a factor in regulating cell proliferation.

*2. Effect of Collagenous Matrices*

One of the most impressive demonstrations of matrix effects is the induction of cartilage, bone, and marrow in animals by dried, devitalized, and demineralized bone matrix powders.[31] S.c. implantion of the bone matrix, which consists largely of Type I collagen, results in a transient inflammatory response, followed by the appearance of mesenchymal stem cells that differentiate into chondrocytes and form cartilage. The resulting cartilage is replaced by bone, and hemopoietic tissue is formed in the bone. This sequence is not unlike that seen in fracture healing. In vitro, the bone powders stimulate mesenchymal cells from muscle to form cartilage.[32] A factor bound to the collagen has been proposed to be the actual "inducer". Other matrices, such as that of neural pigmented epithelium, stimulate the neural crest cells to differentiate into cartilage.[33]

*3. Effect of Collagenase and Proline Analogues*

Addition of bacterial collagenase to developing systems has also been used to assess the role of collagen in differentiation. While some studies may have been carried out with collagenase contaminated with proteases and other enzymes, these data also suggest that collagen has an important role in differentiation. For example, feather morphogenesis, the branching of salivary glands, and the development of the uteric bud are all blocked by collagenase.[34–36] Proline analogues such as *L*-azetidine-2-carboxylic acid which inhibit the hydroxylation and secretion of collagen also inhibit differentiation in some systems.[37] The branching of lung and of salivary glands and ameloblast differentiation are blocked by this compound.

*4. Effect of Fibronectin*

Fibronectin is an adhesive protein which also alters the phenotypic properties of cells. For example, when added to transformed cells, it induces a more normal morphology.[41,42] It also inhibits the pigmentation of neural crest cells,[39] chondrogenic expression of chondrocytes,[40] and, as already discussed, myotube formation.[38]

In summary, collagen and probably other matrix components such as fibronectin can influence the differentiation of a variety of cells, suggesting that these components probably have similar in vivo functions.

**B Growth and Transformation**

Although collagenous substrates were known to promote the growth of primary strains in culture,[6] it has recently been shown that the synthesis of a collagen substrate is necessary for the growth of normal cells even on plastic culture dishes.[11] In these studies, it was found that fibroblasts were "killed" in culture by *cis*-hydroxyproline, a proline analogue. *Cis*-hydroxyproline is incorporated into protein and prevents the formation of the normal *trans*-hydroxyproline in collagen chains. Under these conditions, the protein does not become helical and the deposition of a collagenous matrix is prevented. Fibroblasts can grow normally in the presence of *cis*-hydroxyproline if cultured on collagen.

In an interesting extension of these studies, it was found that *cis*-hydroxyproline also inhibited the growth of mammary epithelial cells.[53] A Type I collagen substrate did not allow cell survival, while a Type IV collagen substrate did. In general, these studies show that *cis*-hydroxyproline is a useful probe for evaluating the participation of specific collagens in the growth of cells in culture. Previous studies suggest that ascorbic acid enhances cell growth, perhaps by its known effect in stimulating collagen synthesis.[54]

Studies with *cis*-hydroxyproline indicate that transformed cells are less dependent on collagen for growth and are much more resistant to *cis*-hydroxyproline than normal cells.[55,56] The molecular basis of the different sensitivity is not clear, although it is not the direct result of a reduced effect of the analogue on collagen synthesis. Rather, it appears that the transformed cells use other mechanisms to attach to culture dishes or require fewer attachment points than normal cells to survive in culture.

The synthesis of collagen is altered in transformed cells.[57,58] After transformation of chick embryo fibroblasts with Rous sarcoma virus, collagen synthesis decreases to one tenth or less. Other cells, such as WI 38 fibroblasts after transformation, synthesize the same amount of procollagen as the nontransformed cells, but it is not converted to collagen or deposited in the matrix.[59] Part of the difference in the morphology and reduced adhesiveness of transformed cells could be a result of reduced matrix proteins. Under some conditions, this can be reversed by factors such as fibronectin[60] and substances that increase the levels of fibronectin and collagen in the cell such as dexamethasone.[61]

Certain metastatic cell lines of murine origin have been examined for attachment to different collagens.[62] The metastatic cells show a marked preference for Type IV collagen in the absence of fibronectin. In the presence of exogenous fibronectin, the cells can also bind to Types I, II, and III collagens. These factors could be important in the metastatic process by allowing the cells to bind to a variety of extracellular matrices and thus survive in different anatomical sites.

### C. Wound Healing

Matrix components have a variety of activities which aid in the repair of a wound (see Figure 2). At the site of a wound, serum provides the fibronectin[63] which can link to fibrin and to collagen and thus stabilize the clot.[64,65] Platelet adherence and aggregation appear to involve both fibronectin and collagen.[66,67] Macrophages also make fibronectin[68] and may utilize it for adhesion to collagen. Fibronectin has opsonic activity for macrophages which could be important for debriding the wound.[69,70] Macrophages release chemotactic factors, including fibronectin, which could stimulate fibroblasts to enter the wound area and begin repair reactions.[71] Collagen[72] and fibronectin[73] are chemotactic and capable of stimulating further deposition of more matrix proteins.[74] Thus, both collagen and fibronectin have key roles at many stages of wound repair.

## III. ROLE OF MATRIX COMPONENTS IN CELL ADHESION

### A. Preparation of Collagen Substrates

A variety of methods exist for the preparation of collagen substrates (see Table 2). The easiest substrate to prepare involves drying an acid solution of collagen onto the culture dish.[75] Such substrates have been successful, for example, in initiating myoblast fusion[7] and in stimulating stromal production by corneal fibroblasts.[9] Generally, collagen is solubilized at 1 mg/mℓ in 0.5 *N* acetic acid in the cold overnight with stirring. The solution is then diluted with water and an aliquot (10 μg/mℓ/35 mm culture dish is adequate to yield an even coat) is allowed to dry on the dish. Various commercial sources of dry and soluble collagen are available. Once dry, the collagen can be sterilized where necessary, by exposure to UV light at a distance of 1 to 2 ft for 6 to 24 hr.

In some cases, hydrated gels of native collagen may be required.[6,8] A 1 mg/mℓ solution of collagen is neutralized either by direct addition of concentrated phosphate buffer to a final concentration of 0.01 *M*, pH 7.4, or by dialysis against the 0.01 *M* phosphate, pH 7.4, at 4°C. The solution is then placed in the petri plate and warmed to 37°C for about 1 hr. A similar gel can be obtained by exposing an acid-solubilized solution of collagen on the plate to ammonia vapors from a $NH_4OH$-soaked cloth in a desiccator for 5 to 10 min.[6,13] The collagen can be sterilized prior to application to the petri dish with gentamicin or by dialysis against buffer saturated with chloroform and subsequent dialysis against buffer to remove the chloroform. The gel

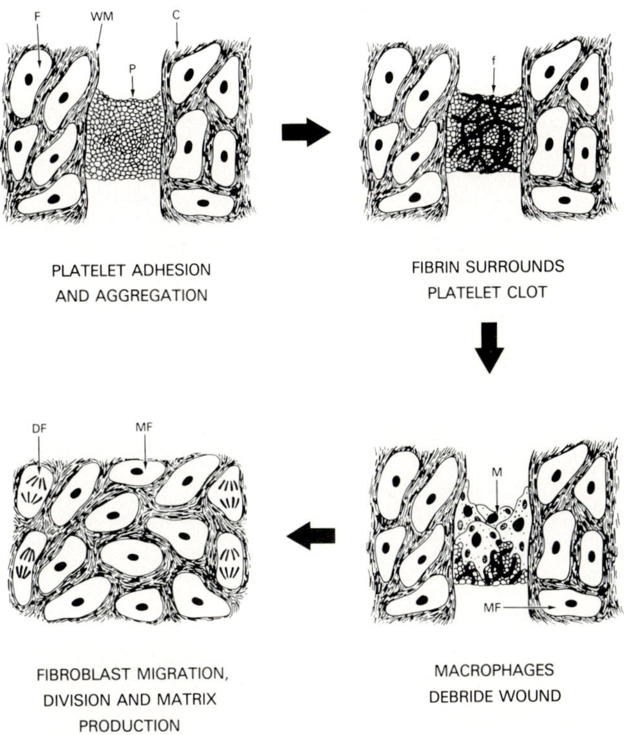


FIGURE 2. Schematic model of the effect of matrix components on wound healing. At the site of a wound in a connective tissue containing fibroblasts (F), platelets (P) adhere and aggregate in response to the exposed collagen (C) at the wound margin (WM). With aggregation, they release their granules which contain fibronectin, growth factors, and enzymes needed for clot formation. As the clot forms, the platelets condense and transglutaminase (Factor $XIII_a$) cross-links the fibrin (f) to fibrin and to fibronectin. Transglutaminase also can cross-link fibronectin to collagen. Macrophages (M) (possibly in response to the fibronectin) bind to the clot and debride the wound. Macrophages are known to make fibronectin and utilize it for opsonization. At this time, the surrounding fibroblasts migrate (MF = migrating cell) into the wound, possibly in response to fibronectin, and begin reestablishing their matrix and proliferating (DF = dividing cell).

can be released from the dish by gentle agitation which results in a floating gel.[76–78] This form has been particularily satisfactory for culture of hepatocytes.[76] Fibroblasts, cultured in this manner, shrink the gel and appear to form a tissue.[77]

The matrix laid down by cells in tissue culture can be used to culture other cells. For example, epithelial and endothelial cells leave a basement membrane which can be used to study epithelial cell function and metastatic tumor cell adhesion.[79,80] In addition, nerve cells leave behind a matrix which promotes the outgrowth of subsequent nerve cell cultures.[81] Many matrix-containing tissues, such as devitalized bone powders,[32] frozen and thawed lens,[9] and crude matrix extracts,[82] have been successfully used to culture cells and promote their differentiation. Probably other matrices can be used to culture and induce differentiation in cells that were previously unable to be cultured or have lost their differentiated function in culture.

**Table 2**
**PREPARATION AND USE OF COLLAGENOUS SUBSTRATES**

| Substrate | Preparation | Use in culture | Ref. |
|---|---|---|---|
| Dried collagen film (10 μg/35 mm dish) | Air dry purified collagen dissolved in 0.5 *N* HAc | Corneal fibroblast stromal production | 9 |
| | | Myoblast fusion | 7 |
| | | Breast epithelial cell culture | 53 |
| | | Epidermal cell culture | 29 |
| | | Fibroblast chemotaxis | 72 |
| Reconstituted collagen gels (0.5–2 mg/35 mm dish) | Expose concentrated collagen (1–2 mg/mℓ) to $NH_4OH$ vapors for 5–10 min or raise pH to neutral with concentrated phosphate to 0.2 *M* | Fibroblast growth | 6,8 |
| | | Platelet adhesion | 67 |
| Floating reconstituted collagen gels (0.5–2 mg/35 mm dish) | Release prepared reconstituted collagen gel mechanically | Hepatocyte culture | 76 |
| | | Fibroblast growth | 77 |
| | | Mammary epithelial cell casein production | 78 |
| Cell matrix | Material remaining on plate after cells are removed with EGTA or detergent | Epithelial cell function | 79 |
| | | Metastatic tumor cell adhesion | 80 |
| | | Nerve cell outgrowth | 81 |
| Tissue material | Decalcified bone powders | Fibroblast and muscle chondrogenesis | 32 |
| | Freeze-thawed lens | Corneal fibroblast stromal production | 9 |

*Note:* The dried substrated can be sterilized if needed by exposure to UV light for 6–20 hr. The reconstituted collagen gels can be prepared in a sterile manner if gentamicin is added or if the collagen is dialyzed against buffer saturated with chloroform followed by subsequent dialysis against buffer alone.

### B. Fibronectin-Mediated Cell Adhesion

Fibronectin mediates the attachment of many cells to collagenous and plastic substrates.[14,47,83] It binds first to collagen and then to the cells in the presence of calcium or magnesium.[13] A great deal is known about the interaction of fibronectin with collagen. In general, native collagen is less active than denatured collagen in binding fibronectin and the following order of activity has been established:[23–26] DENATURED TYPE III > DENATURED TYPE I, II, IV, NATIVE TYPE III > DENATURED TYPE V > NATIVE TYPE I, II, IV. Elastin and *Ascaris* collagen are inactive, while Clq which contains a collagenous segment is active. Only certain segments of fibronectin[84–87] and collagen[23,24] have been found to participate in the interaction of the molecules, thus demonstrating the specificity of the binding.

Attachment of cells to collagen is assessed by the assay outlined in Figure 3.[13,14,27,29,47,52,62,73] In this assay, collagen-coated plates prepared as described in Table 2 are preincubated in culture medium containing various concentrations of

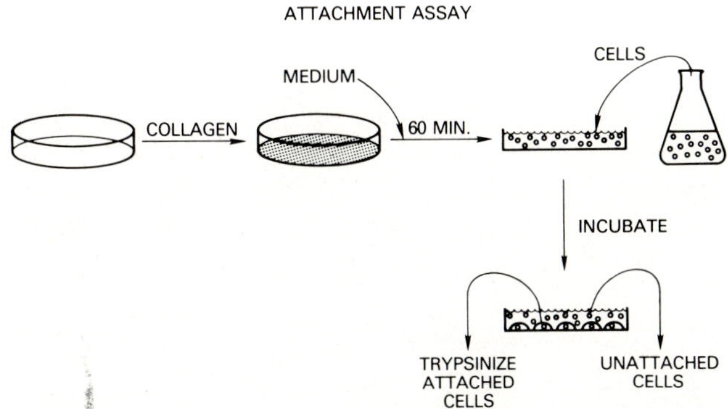


FIGURE 3. Attachment assay. After preparation of the substrate (see Table 2), the freshly isolated cells are added to the dish in the attaching buffer which is generally medium containing 200 μg/mℓ bovine serum albumin and serum where desired. After incubation, usually at 35 to 37°C for the prescribed time, the unattached cells are decanted, the plate washed three times with phosphate buffered saline, and the attached cells are released with trypsin-EDTA (0.1%, 0.1%) and counted electronically.

material being tested for attachment activity (serum, fibronectin, tissue extracts, etc.). The cells are then added and incubated for varying lengths of time, depending on the type of cell. After incubation, the unattached cells are rinsed off and the attached cells are removed following treatment with trypsin/EDTA (see Figure 3 for detailed description). In a typical attachment assay using CHO cells (see Figure 4), the cells adhere and spread in less than 1 hr in the presence of serum or exogenous fibronectin. If fibronectin is not present, the cells adhere more slowly probably utilizing the fibronectin that they synthesize. Using this assay, various cells, including myoblasts,[47] CHO,[13] BHK,[43] skin fibroblasts,[29] rat embryonic periosteum,[88] and hepatocytes,[89] have been found to require fibronectin for adhesion (see Table 3). In general, cells that originate from fibronectin-containing tissues utilize fibronectin for adhesion and adhere well to interstitial collagens. Approximately 1 to 5 μg of fibronectin will mediate the attachment of up to 5 × $10^6$ cell per 35 mm Petri dish coated with 10 μg of collagen.

With fibronectin-dependent cells other macromolecular factors are not needed.[92] Fibronectin, however, does bind both hyaluronic acid and heparin.[93,94] Further, heparin has been shown to increase the binding of fibronectin to collagen.[93,95] These findings suggest that proteoglycans may also participate in maintaining the cells within the matrix.

### C. Nonfibronectin-Mediated Cell Adhesion

#### *1. Chondrocytes*

Cartilage lacks fibronectin[96] and the adhesion of chondrycotyes to collagen does not require fibronectin (see Figure 5).[27] Chondrocytes adhere preferentially to Type II collagen in the presence both of normal serum and of serum from which the fibronectin has been removed. These cells adhere to collagen by means of a different protein, termed chondronectin, which appears to be analogous to fibronectin in function (Figure 6).[27] Chondronectin is also found in chondrocyte-conditioned medium and in extracts of cartilage and of a chondrosarcoma. It is also found in the vitreous fluid which contains Type II collagen, but is not found in extracts of noncollagenous tissues.[108] Chondronectin is more heat labile than fibronectin ($t_{1/2}$ 52 vs.

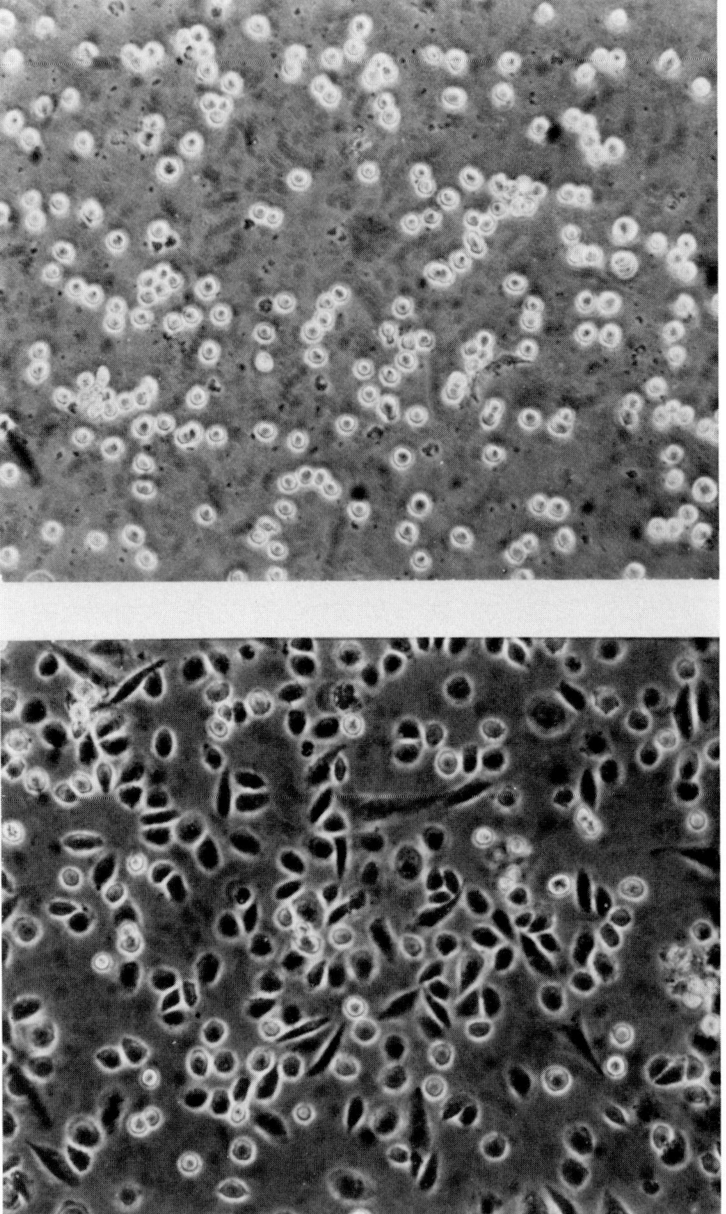

FIGURE 4. Effect of serum on attachment of CHO cells to collagenous matrices. Cells ($1 \times 10^5$) have been allowed to adhere for 1 hr in the presence (bottom) and absence (top) of serum. In the presence of serum, the cells have adhered and are spread. In the absence of serum, the cells are round and floating. Removal of the medium in the latter case would result in complete loss of the cells from the culture dish.

**Table 3**
**ADHESION OF CELLS TO COLLAGEN**

| Cell type | Preferred collagen Substrate | Adhesion factor | Ref. |
|---|---|---|---|
| Established cells lines (CHO, 3T3, etc.) | I–IV | Fibronectin | 13,14 |
| Primary fibroblasts | I–IV | Fibronectin | 29 |
| Rat embryonic periosteum | I–IV | Fibronectin | 88 |
| Rat hepatocytes | I–IV | Fibronectin | 89 |
| Osteosarcoma | I–IV | Fibronectin | 88 |
| Myoblasts | I–IV | Fibronectin | 47 |
| Chick sternal chondrocytes | II | Chondronectin | 27 |
| Breast epithelial cells | IV | ? (not fibronectin) | 53 |
| Guinea pig epidermal cells | IV | laminin | 29,109 |
| EHS sarcoma cells | IV | laminin | 52 |
| TERA (transformed epidermal cells) | IV | laminin | 52 |
| PAM 212 (transformed epidermal cells) | IV | (not fibronectin) laminin | 51 |
| PMT (highly metastatic pulmonary tumor cell) | IV | laminin | 62 |
| Human chorid epithelial cells | IV | laminin | 52 |
| PYS (pareital yolk sac cells) | IV | laminin | 52 |
| Monkey pigmented epithelial cells | IV | laminin | 52 |
| Sheep aorta vascular endothelial cells | IV | laminin | 52 |
| Monkey lens epithelial cells | IV | laminin | 90 |
| Bovine lens epithelial cells | IV | ? | |
| Smooth muscle cells (embryonic sheep aorta) | I, II, III, IV | Fibronectin (for I, II, III), ? for V | 91,97 |

56°C) and of lower molecular weight (180,000 vs. 440,000 daltons). It is active at concentrations as low as 5 to 50 ng/mℓ and thus is more active on a weight basis than fibronectin. A full description of the isolation and properties of chondronectin can be found in Volume 1 of this series.

*2. Epithelial Cells*

A variety of epithelial cells, including breast epithelial cells, guinea pig skin epidermal cells, EHS sarcoma cells, TERA cells, PAM 212 cells, and parietal yolk sac cells, do not synthesize fibronectin and attach preferentially to Type IV collagen (see Table 3, Figure 7).[52] A large glycoprotein, laminin, which is present in all basement membranes, mediates the adhesion of these cells.[28] Laminin binds preferentially to Type IV collagen and shows little activity with Types I, II, III, and V collagen substrates. Its role in cell adhesion has been further confirmed by the finding that antibodies directed against laminin block epithelial cell adhesion, while antibodies against fibronectin have no effect.

*3. Direct Binding of Cells to Collagen Substrates*

***a. Smooth Muscle Cells***

Smooth muscle cells adhere preferentially to Type V collagen in the absence of fibronectin or serum (see Figure 6).[91] Fibronectin, however, will stimulate their adhesion to collagens I, II, and III.[91,97] It appears that these cells adhere to Type V collagen directly without the need of additional adhesion glycoproteins, since cells pretreated with cycloheximide attach as well as the untreated counterparts to Type

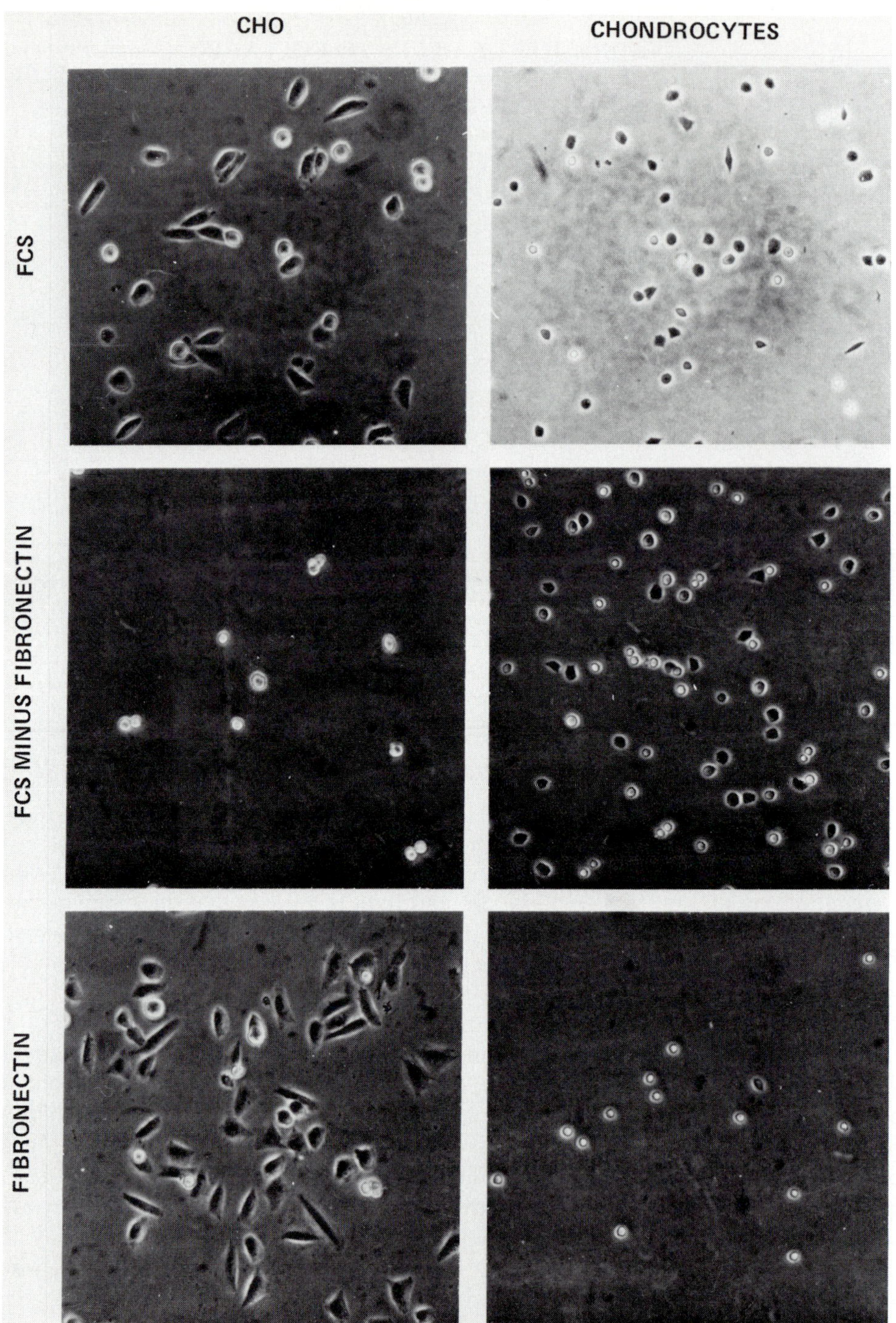


FIGURE 5. Attachment of CHO cells and chondrocytes in the presence of serum, serum depleted of fibronectin, and pure fibronectin. CHO cells and chondrocytes ($1 \times 10^5$) were allowed to attach either in the presence of 10% fetal calf serum, 10% fetal calf serum depleted of fibronectin, or approximately 20 μg of pure fibronectin. CHO cells adhered in the presence of serum and fibronectin but not in the presence of serum depleted of fibronectin. Chondrocytes adhered in the presence of serum and serum depleted of fibronectin, but not in the presence of fibronectin. All assays were carried out on Type II collagen, and the unattached cells were removed before the picture was taken.

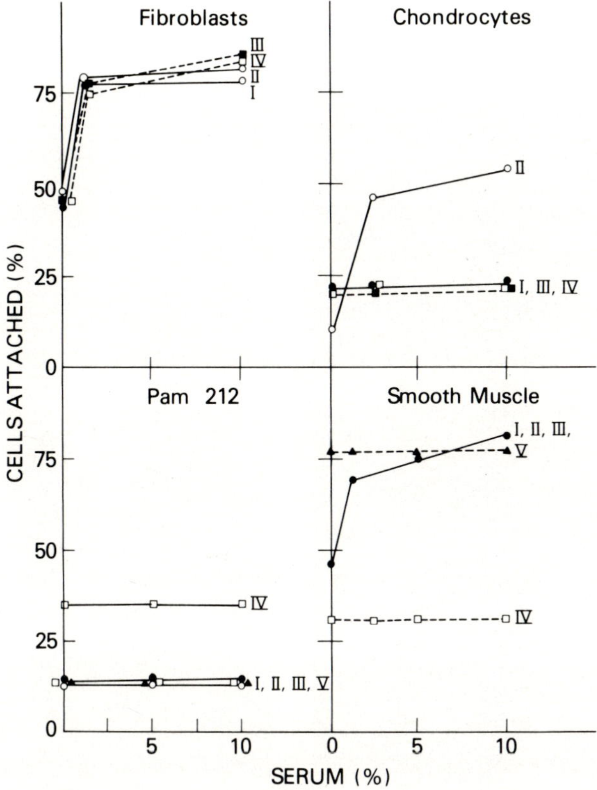


FIGURE 6. Attachment activities of fibroblasts, chondrocytes, PAM 212 cells, and smooth muscle cells in the presence of increasing amounts of serum on collagens I, II, III, IV, and V. All assays were carried out for 2 hr on collagens prepared by established procedures. Human skin fibroblasts were obtained from American Type Culture (Rockville, Md.), chondrocytes were obtained from embryonic chick sternum, PAM 212 was obtained from S. Yuspa (National Cancer Institute), and smooth muscle cells were obtained from embryonic sheep aorta.

V collagen. In addition, soluble Type V collagen will inhibit smooth muscle cell binding to Type V collagen substrates, while soluble Type I collagen is without effect. Laminin and chondronectin are not effective with these cells. These data suggest that smooth muscle cells bind directly to Type V collagen via a membrane component. The Type V collagen receptor is not trypsin-sensitive, but appears to contain sugar residues since both neuraminidase and Concavalin A treatment of the cells will inhibit attachment to Type V collagen.

***b. Platelets***

Platelets contain fibronectin and release it during their aggregation by collagen or thrombin.[98] While it has been suggested that fibronectin on the platelet cell surface acts as the collagen receptor,[66] this is not well established.[67] However, once platelets adhere to collagen, exogenous fibronectin enhances their flattening.[99] Further evidence of a role for fibronectin in platelet function is suggested by studies on a patient with Ehlers-Danlos Syndrome who exhibited defective platelet aggregation and poor wound healing.[100] Using an immunological assay, it was found that the levels of

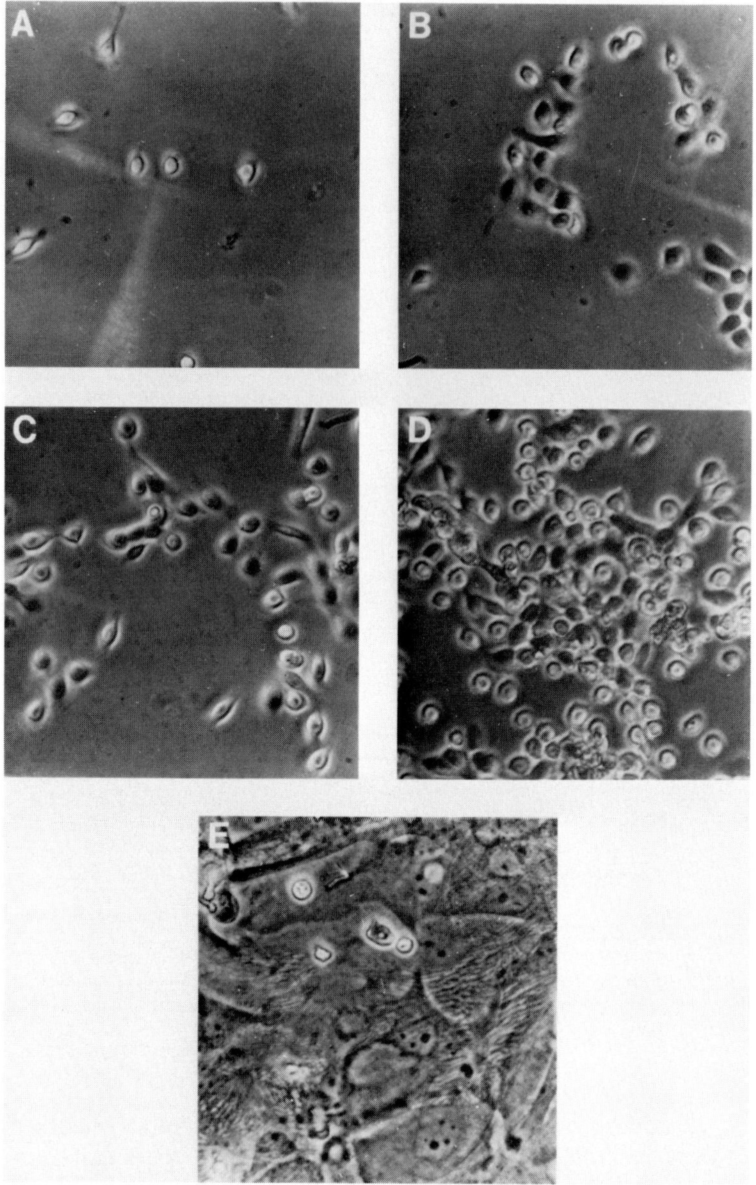


FIGURE 7. Attachment and growth of guinea pig epidermal cells on collagens I, II, III, and IV. After an 18-hr incubation in the absence of serum, the unattached cells were removed and the attached cells on collagens I (A), II (B), III (C), and IV (D) are shown. After 5 days in culture, the adherent cells on Type IV collagen (E) have differentiated into multilayered squamous epithelium.

fibronectin in the patient's blood were normal, but that her platelets aggregated poorly in the presence of her own plasma. However, in the presence of plasma from a control subject, her platelets aggregated normally. Addition of purified fibronectin from a control subject to her plasma also restored platelet aggregation. These data suggest that the fibronectin from this patient is abnormal and does not function well in platelet aggregation. An abnormal fibronectin may also have contributed to her joint hypermobility and poor wound healing.

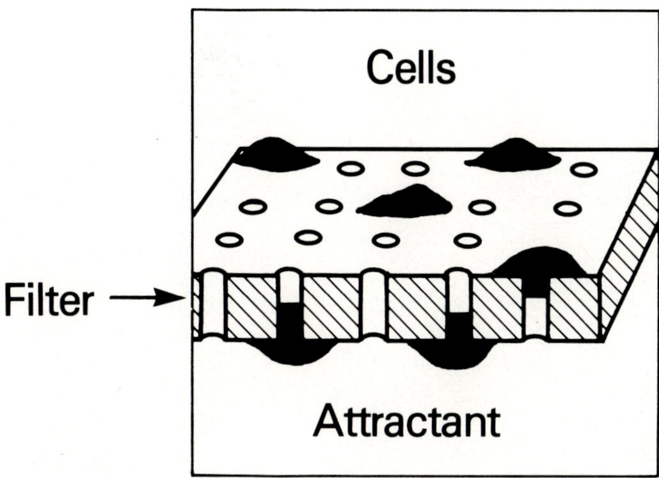


FIGURE 8. Schematic model of a Boyden chamber. A porous filter (8-μm pore size) separates the lower compartment which holds the attractant from the upper chamber to which the cells are added. After 1 to 4 hr, cells have migrated through the filter and can be examined by the light microscope.

### D. Methods for Determining Matrix Requirements for Cell Adhesion

The adhesion characteristics of cells can be measured and the conditions modified to maximize cell adhesion. In the attachment assay, variables, such as the time required for adhesion, ion concentration, preparation of the cells, and the presence of an appropriate substrate and cofactor molecules, must be tested to optimize the attachment. The source and isolation procedures used for the cells as well as enzymatic pretreatment of the cells can influence the adhesion properties. For example, the fibronectin receptor on fibroblasts is not trypsin-sensitive,[13] while the collagen receptor on hepatocytes is.[101] In fact, depending on how hepatocytes are isolated, they may either bind directly to collagen[102] or utilize fibronectin to adhere.[89] The nature of the component on the cell surface which participates in adhesion can be further evaluated by lectin pretreatment of the cells. For example, smooth muscle cells pretreated with Concanavalin A will not bind to collagen substrates, whereas, pretreatment of the substrate has no effect.[91] Thus, the interacting molecule on the cell surface is a glycoconjugate.

Various substrates, such as collagens I to V, tissue extracts, and the matrix deposited by the cells in culture, can be tested. For example, fibroblasts adhere well to all collagens in the presence of serum,[29] while chondrocytes adhere preferentially to Type II collagen[27] and epidermal cells to Type IV collagen.[29] The requirement for additional factors can be tested by reconstitution assays. The effect of serum, fibronectin-free serum, or purified fibronectin on cell adhesion can be used to evaulate whether fibronectin is necessary for cell adhesion. Although fibroblasts synthesize fibronectin, they require exogenous fibronectin for attachment when protein synthesis is inhibited with cyclohexamide.[92] Under the same conditions, chondrocytes utilize a different serum protein and fibronectin does not stimulate their attachment.[27] The attachment of certain epithelial cells is not stimulated by serum.[52] Smooth muscle cells attach to Type I collagen via fibronectin,[97] while the attachment to Type V collagen is unaffected by serum or by fibronectin.[91]

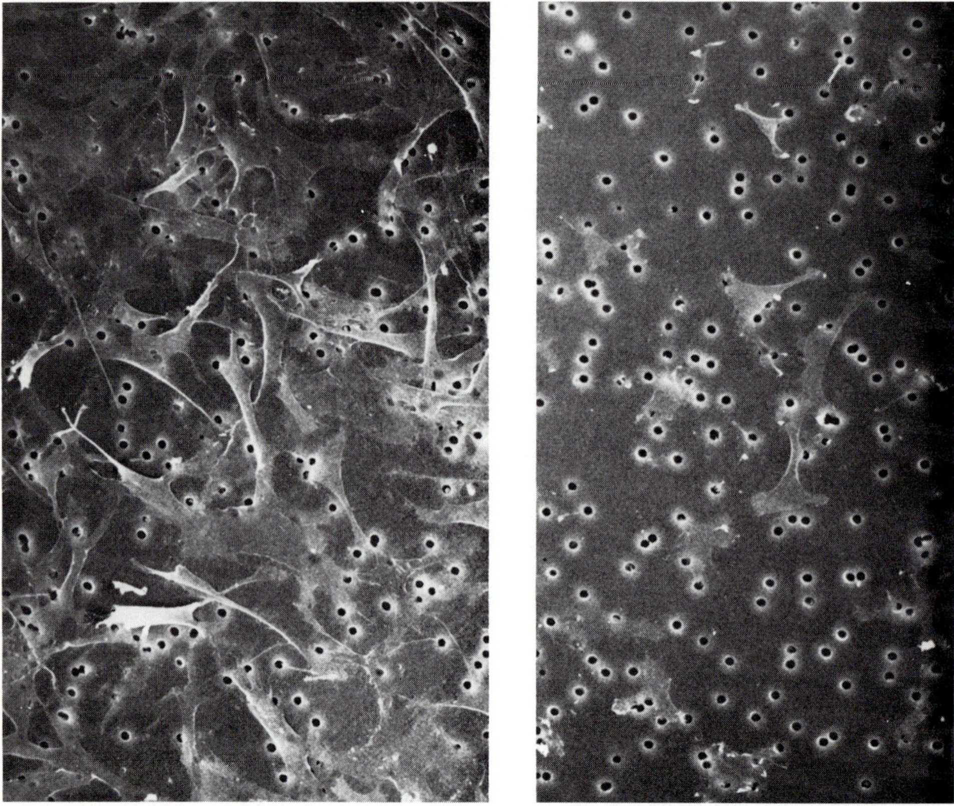

FIGURE 9. Appearance of fibroblasts in scanning electron microscopy after migration through the filter in the Boyden chamber in the presence (left) and absence (right) of fibronectin. (Prepared by M. Silver (National Institute of Dental Research.)

Pretreatment of the matrix or cells with an adhesion factor, followed by a wash to remove unbound material, may aid in identifying the mechanism by which the molecule works. For example, fibronectin preincubated for 1 hr with the cells followed by a wash does not enhance adhesion.[13] However, if the collagen substrate is preincubated with fibronectin for the same length of time and then washed, the cells adhere well. Thus, fibronectin acts by binding first to the substratum. Laminin works in a similar manner,[52] but chondronectin has no activity when the substrate is pretreated for 1 hr and only moderate activity when the cells are preincubated.[27] The identity of an adhesion factor can be confirmed by treating the culture with antibody against the adhesion molecule and monitoring its ability to block attachment.[52,92]

In general, cells that adhere slowly in the presence or absence of serum are probably making their own adhesion factors. Synthesis of required molecules by the cells can be confirmed by pretreatment of the cells with protein synthesis inhibitors such as cycloheximide for 3 hr before and during the assay at a concentration of 5 to 25 μg/mℓ. Such an approach was successful in determining the adhesion requirements of the PAM 212 cells.[52] These cells adhered slowly, requiring longer than 6 hr for maximal attachment. In the presence of cycloheximide, no attachment occurred. However, crude extracts of basement membrane tissues and later pure laminin were found to stimulate maximal cell attachment within 1 hr, even in the

**Table 4**
**EFFECT OF FIBRONECTIN AND FRAGMENTS FROM FIBRONECTIN ON FIBROBLAST CHEMOTAXIS**

| Substance tested | Chemotactic response (cell/mm$^2$) |
|---|---|
| None | 73 ± 16 |
| Plasma fibronectin | 275 ± 36 |
| Cell surface fibronectin | 268 ± 14 |
| Collagen binding domain of fibronectin | 14 ± 7 |
| Cell binding domain of fibronectin | 214 ± 36 |

*Note:* The attractants, each at 20 μg/mℓ in Dulbecco's minimal essential medium, were placed in the lower well of a "blind well" Boyden chamber. The gelatin-coated polycarbonate filters (Nucleopore)® were then placed above the fibronectin solutions, and the upper chambers were filled with a suspension of human skin fibroblasts (300,000/mℓ). After incubation at 37°C for 4 hr, the filter was removed and the cells were fixed and stained. The number of cell nuclei on the lower surface were counted under a microscope. Plasma fibronectin was preapred by gelatin-Sepharose® affinity chromatography.[69] Cell surface fibronectin and its fragments were prepared as previously described.[85]

absence of protein synthesis. It should be mentioned that not all cells in a cell population may be able to adhere. For example, up to 80 to 90% of the skin fibroblasts will adhere in an assay, while only 20% of guinea pig epidermal cells attach.[29] Epidermal cells adhere less well because the epidermis is multilayered and only a fraction of the cells (i.e., the basal cells) are in contact with the basement membrane. It has been demonstrated that the adherent cells in vitro are derived from the basal layer of cells which touches the basement membrane in vivo.[103]

In general, it has been found that cells interact with extracellular matrix components present in the tissues from which they are derived. Detailed characterization of the requirements for cell adhesion should lead to improved efficiency of culture. Since many matrix molecules promote cell differentiation, their presence during culture should help to maintain and enhance the specialized functions of cells.

## IV. ROLE OF ADHESION AND MATRIX MOLECULES IN CHEMOTAXIS

The ability of cells to detect and move toward various diffusable substances is known as chemotaxis and is important in tissue formation and wound repair. The migration of cells is usually studied in a Boyden chamber, a cylinder containing a porous filter which divides the cylinder into an upper and lower compartment (see Figure 8). Cells are generally placed in the upper compartment while the attractant is placed in the lower one. After a suitable time (usually 2 to 4 hr), the cells that have migrated through the pores of the filter (see Figure 9) can be quantitated by direct counting of the cells on the bottom side of the filter.

Using the Boyden chamber assay, the migration of macrophages and leucocytes has been characterized.[104] Recently, the migration of connective tissue cells has been studied.[72,73] However, unlike the circulating cells, fibroblasts would not adhere directly to the filters used in the assay.[105] Since cells cannot migrate unless they adhere, collagen coating the filters was a prerequisite for the fibroblasts to adhere and migrate. In this system, collagen is also chemotactic.[72] Fibronectin has been shown to stimulate both random migration of cultured cells[106] and directed movement in the Boyden chamber (see Table 4).[73] By varying the amount of fibronectin above and below the filter in the so called "checkerboard" analysis, fibronectin has been demonstrated to be a true chemoattractant. Using fragments of fibronectin produced by proteases, it has been found that the cell-binding portion of the fibronectin molecule retains the chemoattractant activity (see Table 4), while the collagen-binding region is inactive. Thus, it is a specific region of the fibronectin molecule which stimulates fibroblast migration and in a wound could enhance repair. Preliminary studies with smooth muscle cells also demonstrate that fibronectin is chemotactic for cells plated on Type I collagen while it is without effect on Type V collagen.[91]

Thus, definition of the adhesion requirements of cells will allow further exploration of mechanisms of cell migration involved in wound repair, development (such as the neural crest), disease, and metastases.

## V. SUMMARY

Extracellular matrices are composed of macromolecules which are unique to each tissue and are capable of influencing the function of cells within the tissues. Matrices not only provide structural support to tissues, but also determine the cells that populate the tissues. Matrix components also have biological activities, such as promoting cell growth, differentiation, and migration. Cell-collagen interactions are mediated via specific glycoproteins and are important in the adhesion, growth, differentiation, and transformation of cells. Alterations in the interaction of cells with their matrix could result in tissue malfunction and malformation and underly various health problems, such as poor wound healing, abnormal development, fibrotic disease, and the metastatic spread of tumor cells.

## REFERENCES

1. **Gallop, P. M., Blumenfield, O. O., and Seifter, S.,** Structure and metabolism of connective tissue proteins, *Annu. Rev. Biochem.*, 41, 617, 1972.
2. **Miller, E. J. and Matukas, V. J.,** Biosynthesis of collagen: the biochemist's view, *Fed. Proc. Fed. Am. Soc. Exp. Biol.*, 33, 1197, 1974.
3. **Martin, G. R., Byers, P. H., and Piez, K. A.,** Procollagen, *Adv. Enzymol.*, 42, 167, 1975.
4. **Ramachandran, G. N. and Reddi, A. H.,** *Biochemistry of Collagen*, Plenum Press, New York, 1976.
5. **Weiss, P.,** Experiments on cell and axon orientation in vitro: the role of colloidal exudates in tissue organization, *J. Exp. Zool.*, 100, 353, 1945.
6. **Ehrmann, R. L. and Gey, G. D.,** The growth of cells on a transparent gel of reconstituted rat-tail collagen, *J. Natl. Cancer Inst.*, 16, 1375, 1956.
7. **Hauschka, S. D. and Konigsberg, I.,** The influence of collagen on the development of muscle clones, *Proc. Natl. Acad. Sci. U.S.A.*, 55, 119, 1966.
8. **Elsdale, T. and Bard, J.,** Collagen substrate for studies on cell behavior, *J. Cell Biol.*, 54, 626, 1972.
9. **Meier, S. and Hay, E. D.,** Control of corneal differentiation by extracellular materials. Collagen as a promoter and stabilizer of epithelial stroma production, *Dev. Biol.*, 38, 249, 1974.
10. **Emerman, J. T. and Pitelka, D. R.,** Maintenance and induction of morphological differentiation in disassociated mammary epithelium cells on floating collagen membranes, *In Vitro* 13, 316, 1977.

11. **Liotta, L. A., Vembu, D., Kleinman, H. K., Martin, G. R., and Boone, C.**, Collagen required for proliferation of cultured connective tissue cells but not their transformed counterparts, *Nature*, 272, 622, 1978.
12. **Gospodarowicz, D., Greenburg, G., and Birdwell, C. R.**, Determination of cellular shape by the extracellular matrix and its correlation with the control of cellular growth, *Cancer Res.*, 38, 4155, 1978.
13. **Klebe, R. J.**, Isolation of a collagen-dependent cell attachment factor, *Nature*, 250, 248, 1974.
14. **Pearlstein, E.**, Plasma membrane glycoprotein which mediates adhesion of fibroblasts to collagen, *Nature*, 262, 497, 1976.
15. **Schor, A. M., Schor, S. L., and Kumar, S.**, Importance of a collagen substratum for stimulation of capillary endothelial cell proliferation for tumor angiogenesis factor, *Int. J. Cancer*, 24, 225, 1979.
16. **Kleinman, H. K., Hewitt, A. T., Murray, J. C., Liotta, L. A., Rennard, S. I., Pennypacker, J. P., McGoodwin, E. B., Martin, G. R., and Fishman, P. H.**, Cellular and molecular specificity in the interaction of adhesion proteins with collagen and with cells, *J. Supramol. Struct.*, 11, 69, 1979.
17. **Yaoita, H., Foidart, J. M., and Katz, S. I.**, Localization of the collagenous component in skin basement membrane, *J. Invest. Dermatol.*, 70, 191, 1978.
18. **Kanwar, Y. S. and Farquhar, M. G.**, Isolation of glycosaminoglycans (heparan sulfate) from glomerular basement membranes, *Proc. Natl. Acad. Sci. U.S.A.*, 76, 4493, 1979.
19. **Hassell, J. R., Robey, P. G., Barrach, H.-J., Wilczek, J., Rennard, S. I., and Martin, G. R.**, Isolation of a heparan sulfate containing proteoglycan from basement membrane, *Proc. Natl. Acad. Sci. U.S.A.*, 77, 4494, 1980.
20. **Vacarro, C. A. and Brody, J. B.**, Ultrastructural localization and characterization of proteoglycans in the pulmonary alveolus, *Am. Rev. Respir. Dis.*, 120, 901, 1979.
21. **Wood, G. C.**, The formation of fibrils from collagen solutions, *Biochem. J.*, 75, 605, 1960.
22. **Mathews, M. B. and Decker, L.**, The effect of acid mucopolysaccharide and acid mucopolysaccharide-proteins on fibril formation from collagen solutions, *Biochem. J.*, 109, 517, 1968.
23. **Dessau, W., Adelmann, B. C., Timpl, R., and Martin, G. R.**, Identification of the sites in collagen α-chains that bind serum anti-gelatin factor (cold insoluble globulin), *Biochem. J.*, 169, 55, 1978.
24. **Kleinman, H. K., McGoodwin, E. B., Martin, G. R., Klebe, R. J., Fietzek, P. P., and Woolley, D. E.**, Localization of the binding site for cell attachment in the α1(I) chain of collagen, *J. Biol. Chem.*, 253, 5642, 1978.
25. **Engvall, E., Ruoslahti, E., and Miller, E. J.**, Affinity of fibronectin to collagen of different genetic types and to fibrinogen, *J. Exp. Med.*, 147, 1584, 1978.
26. **Jilek, F. and Hörmann, H.**, Cold-insoluble globulin (fibronectin) IV. Affinity to soluble collagen of various types, *Hoppe-Seylers Z. Physiol. Chem.*, 359, 247, 1978.
27. **Hewitt, A. T., Kleinman, H. K., Pennypacker, J. P., and Martin, G. R.**, Identification of an adhesion factor for chondrocytes, *Proc. Natl. Acad. Sci. U.S.A.*, 77, 385, 1980.
28. **Timpl, R., Rohde, H., Gehron Robey, P., Rennard, S. I., Foidart, J. M., and Martin, G. R.**, Laminin — a glycoprotein from basement membrane, *J. Biol. Chem.*, 254, 9933, 1979.
29. **Murray, J. C., Stingl, G., Kleinman, H. K., Martin, G. R., and Katz, S. I.**, Epidermal cells adhere preferentially to type IV (basement membrane) collagen, *J. Cell Biol.*, 80, 197, 1979.
30. **Kosher, R. A. and Church, R. L.**, Stimulation of somite chondrogenesis by procollagen and collagen, *Nature*, 258, 337, 1975.
31. **Reddi, A. H. and Anderson, W. A.**, Collagenous bone matrix-induced endochondral ossification and hemopoiesis, *J. Cell Biol.*, 69, 557, 1976.
32. **Nakagawa, M. and Urist, M. R.**, Chondrogenesis in tissue culture under the influence of a diffusable component of bone matrix. *Proc. Soc. Exp. Biol.*, 154, 568, 1977.
33. **Newsome, D. A.**, In vitro stimulation of cartilage in embryonic chick neural cells by products of retinal pigmented epithelium, *Dev. Biol.*, 49, 496, 1976.
34. **Stuart, E. S., Barber, B., and Moscona, A. A.**, An analysis of feather germ formation in the embryo and *in vitro*, in normal development and in skin treated with hydrocortisone, *J. Exp. Zool.*, 197, 97, 1972.
35. **Grobstein, C. and Cohen, J.**, Collagenase: effect on the morphogenesis of embryonic salivary epithelium *in vitro*, *Science*, 150, 626, 1965.
36. **Wessels, N. K. and Cohen, J. H.**, Effects of collagenase on developing epithelia *in vitro*: lung, uteric bud and pancreas, *Dev. Biol.*, 18, 294, 1968.
37. **Alescio, T.**, Effect of a proline analogue, azetidine-2-carboxylic acid, on the morphogenesis in vitro of mouse embryonic lung, *J. Embryol. Exp. Morphol.*, 29, 439, 1973.
38. **Podleski, T. R., Greenberg, I., Schlessinger, J., and Yamada, K. M.**, Fibronectin delays the fusion of $L_6$ myoblasts, *Exp. Cell Res.*, 122, 317, 1979.

39. **Sieber-Blum, M., Sieber, F., Yamada, K. M., and Cohen, A.,** Cell surface fibronectin (FN) in cultures of differentiating neural crest cells, *J. Cell Biol.,* 79, 31a, 1979.
40. **Pennypacker, J. P., Hassell, J. R., Yamada, K. M., and Pratt, R. M.,** The influence of an adhesion cell surface protein on chondrogenic expression *in vitro, Exp. Cell Res.,* 121, 411, 1979.
41. **Yamada, K. M. and Olden, K.,** Fibronectins — adhesive glycoproteins of cell surface and blood, *Nature,* 275, 179, 1978.
42. **Vaheri, A. and Mosher, D. F.,** High molecular weight cell surface glycoprotein (fibronectin) lost in malignant transformation, *Biochim. Biophys. Acta,* 516, 1, 1978.
43. **Grinnell, F.,** Cellular adhesiveness and extracellular substrata, *Int. Rev. Cytol.,* 53, 65, 1978.
44. **Ketley, J. N., Orkin, R. W., and Martin, G. R.,** Collagen in developing chick muscle *in vivo* and *in vitro, Exp. Cell Res.,* 99, 261, 1976.
45. **Hauschka, S. D. and White, N. K.,** Studies on myogenesis *in vitro,* in *Research in Muscle Development and the Muscle Spindle,* Banker, B. O., Pryzbylski, R. J., van der Meulen, J. P., and Victor, M., Eds., Excerpta Medica, Amsterdam, 1972, 53.
46. **Furcht, L. T., Mosher, D. F., and Wendelschaeffer-Craab, G.,** Immunocytochemical localization of fibronectin (LETS proteins) on the surface of $L_6$ myoblasts: light and electron-microscopic studies, *Cell,* 13, 263, 1978.
47. **Chicquet, M., Puri, E. C., and Turner, D. C.,** Fibronectin mediates attachment of chicken myoblasts to a gelatin-coated substratum, *J. Biol. Chem.,* 254, 5475, 1979.
48. **Takasu, N., Charrier, B., Mauchamp, J., and Lissitzky, S.,** Effect of gelatin on the cyclic AMP response of primocultured hog thyroid cells to acute thyrotropin stimulation, *Biochim. Biophys. Acta,* 587, 507, 1979.
49. **Green, H.,** Terminal differentiation of cultured human epidermal cells, *Cell,* 11, 405, 1977.
50. **Lui, S. C. and Karasek, M.,** Isolation and growth of adult human epidermal keratinocytes in cell culture, *J. Invest. Dermatol.,* 71, 157, 1978.
51. **Rheinwald, J. G. and Green, H.,** Serial cultivation of strains of human epidermal keratinocytes: the formation of keratinizing colonies from single cells, *Cell,* 6, 331, 1975.
52. **Terranova, V. P., Rohrbach, D. H., and Martin, G. R.,** Role of laminin in the attachment of PAM 212 (epithelial cells) to basement membrane collagen, *Cell,* 22, 719, 1980.
53. **Wicha, M., Liotta, L. A., Garbisa, S., and Kidwell, W. R.,** Basement membrane collagen requirements for attachment and growth of mammary epithelium, *Exp. Cell Res.,* 124, 181, 1979.
54. **Rowe, D. W., Starman, B. J., Fujimoto, W. Y., and Williams, R. H.,** Differences in growth response to hydrocortisone and ascorbic acid by human diploid fibroblasts, *In Vitro,* 13, 824, 1977.
55. **Kao, W.-Y and Prockop, D. J.,** Proline analogue removes fibroblasts from cultures with mixed cell populations, *Nature,* 266, 63, 1977.
56. **Vembu, D., Liotta, L. A., Paranjpe, M., and Boone, C. W.,** Correlation of tumorigenicity with resistance to growth inhibition by *cis*-hydroxyproline, *Exp. Cell Res.,* 124, 247, 1979.
57. **Kamine, J. and Rubin, H.,** Coordinate control of collagen synthesis and cell growth in chick embryo fibroblasts and the effect of viral transformation on collagen synthesis, *J. Cell. Physiol.,* 92, 1, 1977.
58. **Adams, S. E., Sobel, M. E., Howard, B. H., Olden, K., Yamada, K. M., DeCrombrugghe, B., and Pastan, I.,** Levels of translatable mRNAs for cell surface protein, collagen precursors, and two membrane proteins are altered in Rous sarcoma virus-transformed chick embryo fibroblasts, *Proc. Natl. Acad. Sci. U.S.A.,* 74, 3399, 1977.
59. **Arbogast, B. W., Yoshimura, M., Kefalides, N., Holtzer, H., and Kaji, A.,** Failure of cultured chick embryo fibroblasts to incorporate collagen into their extracellular matrix when transformed by Rous sarcoma virus, *J. Biol. Chem.,*252, 8863, 1977.
60. **Yamada, K. M., Yamada, S. S., and Pastan, I.,** Cell surface protein partially restores morphology, adhesiveness and contact inhibition of movement to transformed fibroblasts, *Proc. Natl. Acad. Sci. U.S.A.,* 73, 1211, 1977.
61. **Furcht, L. T., Mosher, D. F., Wendelschaefer-Crabb, G., and Foidart, J. M.,** Reversal by glucocorticoid hormone of the loss of fibronectin and procollagen matrix around transformed cells, *Cancer Res.,* 39, 2077, 1979.
62. **Murray, J. C., Liotta, L., Rennard, S. I., and Martin, G. R.,** Adhesion characteristics of murine metastatic and nonmetastatic tumor cells *in vitro, Cancer Res.,* 40, 347, 1980.
63. **Mosesson, M. W. and Umfleet, R. A.,** The cold-insoluble globulin of human plasma. I. Purification, primary characterization and relationship to fibrinogen and other cold-insoluble fraction components, *J. Biol. Chem.,* 245, 5728, 1970.
64. **Mosher, D. F.,** Cross-linking of cold insoluble globulin by fibrin-stabilizing factor, *J. Biol. Chem.,* 245, 5728, 1975.

65. **Mosher, D. F., Schad, P. E., and Kleinman, H. K.,** Cross-linking of fibronectin to collagen by blood coagulation factor $XIII_a$, *J. Clin. Invest.*, 64, 781, 1979.
66. **Bensusan, H. B., Kah, T. L., Henry, K. G., Murray, B. A., and Culp, L. A.,** Evidence that fibronectin is the collagen receptor on platelet membranes, *Proc. Natl. Acad. Sci. U.S.A.*, 75, 5864, 1978.
67. **Santoro, S. A. and Cunningham, L. W.,** Fibronectin and the multiple interaction model for platelet-collagen adhesion, *Proc. Natl. Acad. Sci. U.S.A.*, 76, 2644, 1979.
68. **Alitalo, K., Hovi, T., and Vaheri, A.,** Fibronectin is produced by human macrophages, *J. Exp. Med.*, 151, 602, 1980.
69. **Hopper, K. E., Adelmann, B. C., Gentner, G., and Gay, S.,** Recognition by guinea-pig peritoneal exudate cells of conformationally different states of the collagen molecule, *Immunology*, 30, 249, 1976.
70. **Saba, T. M., Blumenstock, F. A., Weber, P., and Kaplan, E.,** Physiological role of cold-insoluble globulin in host defense: implications of its characterization as an $A_2$ surface binding glycoprotein, *Ann. N.Y. Acad. Sci.*, 312, 43, 1978.
71. **Tsukamoto, Y., McCarthy, J. B., and Wahl, S. M.,** Macrophage derived fibroblast chemotactic factor, *J. Dental Res.*, 59, 945, 1980.
72. **Postlethwaite, A. E., Seyer, J. M., and Kang, A. H.,** Chemotactic attraction of human fibroblasts to type I, II, and III collagens and collagen-derived peptide, *Proc. Natl. Acad. Sci. U.S.A.*, 75, 871, 1978.
73. **Gauss-Müller, V., Kleinman, H. K., Martin, G. R., and Schiffmann, E.,** Role of attachment and attractants in fibroblast chemotaxis, *J. Lab. Clin. Med.*, 96, 1071, 1980.
74. **Foidart, J. M., Berman, J. J., Paglia, L., Rennard, S. I., Abe, S., Perantoni, A., and Martin, G. R.,** Synthesis of fibronectin, laminin and several collagens by a liver-derived epithelial line, *Lab. Invest.*, 42, 525, 1980.
75. **Kleinman, H. K., McGoodwin, E. B., Rennard, S. I., and Martin, G. R.,** Preparation of collagen substrates for cell attachment: effect of collagen concentration and phosphate buffer, *Anal. Biochem.*, 94, 308, 1979.
76. **Sattler, G. A., Michalopoulos, G., Sattler, G. L., and Pitot, H. C.,** Ultra structure of adult rat hepatocytes cultured on floating collagen membranes, *Cancer Res.*, 38, 1539, 1978.
77. **Bell, E., Ivarsson, B., and Merritt, C.,** Production of a tissue-like structure by contraction of collagen lattices by human fibroblasts of different potential in vitro, *Proc. Natl. Acad. Sci. U.S.A.*, 76, 1274, 1979.
78. **Emerman, J. T., Burwen, S. J., and Pitelka, D. R.,** Substrate properties influencing ultrastructural differentiation of mammary epithelial cells in tissue culture, *Cell*, 11, 109, 1979.
79. **Vlodavsky, I., Lui, G. M., and Gospodarowicz, D.,** Morphological appearance, growth behavior and migratory activity of human tumor cells maintained on extracellular matrix vs. plastic, *Cell*, 19, 607, 1980.
80. **Poste, G. and Fidler, I. J.,** The pathogenesis of cancer metastases, *Nature*, 283, 139, 1980.
81. **Mizel, S. D. B. and Bamburg, J. R.,** Studies on the action of nerve growth factor. III. Role of RNA and protein synthesis in the process of neurite outgrowth, *Dev. Biol.*, 49, 20, 1976.
82. **Reid, L. M. and Rojkind, M.,** New techniques for culturing differentiated cells: Reconstituted basement membrane rafts, in *Methods in Enzymology*, Vol. 58, Jakoby, W. B. and Pastan, I., Eds., Academic Press, New York, 1979, 263.
83. **Kleinman, H. K., Murray, J. C., McGoodwin, E. B., and Martin, G. R.,** Connective tissue structure: cell binding to collagen, *J. Invest. Dermatol.*, 71, 9, 1978.
84. **Balian, G., Glick, E. M., Crouch, E., Davidson, J. M., and Bornstein, P.,** Isolation of a collagen binding fragment from fibronectin and cold-insoluble globulin, *J. Biol. Chem.*, 254, 1429, 1979.
85. **Hahn, L. E. and Yamada, K. M.,** Identification and isolation of a collagen-binding fragment of the adhesive glycoprotein fibronectin, *Proc. Natl. Acad. Sci. U.S.A.*, 55, 119, 1979.
86. **Ruoslahti, E. and Hayman, E. G.,** Two active sites with different characteristics in fibronectin, *FEBS Lett.*, 97, 221, 1979.
87. **Gold, L. I., Garcia-Pardo, A., Frangione, B., Franklin, E. C., and Pearlstein, E.,** Subtilisin and cyanogen bromide cleavage products that retain gelatin-binding activity, *Proc. Natl. Acad. Sci. U.S.A.*, 76, 4803, 1980.
88. **Kleinman, H. K., Murray, J. C., McGoodwin, E. B., Martin, G. R., and Binderman, I.,** Attachment of bone cells to collagen, in *Proceedings, Mechanisms of Localized Bone Loss*, Horton, J. E., Tarpley, T. M., and Davis, W. F., Eds., *Calcif. Tiss. Abs.*, Spec. Suppl., 1978, 61.
89. **Berman, M. D., Waggoner, J. G., Foidart, J. M., and Kleinman, H. K.,** Attachment to collagen by isolated hepatocytes from rats with induced hepatic fibrosis, *J. Lab. Clin. Med.*, 95, 660, 1980.

90. **Hughes, R. C., Mills, G., Courtois, Y., and Tassin, J.**, Role of fibronectin in the adhesiveness of bovine lens epithelial cells, *Biol. Cell.*, 36, 321, 1979.
91. **Grotendorst, G. R., Seppä, H., Kleinman, H. K., and Martin, G. R.**, Attachment of smooth muscle cells to collagen and their migration to platelet-derived growth factor, *Proc. Natl. Acad. Sci. U.S.A.*, 78, 3669, 1981.
92. **Pearlstein, E. and Gold, L. I.**, High-molecular weight glycoprotein as mediator of cellular adhesion, *Ann. N.Y. Acad. Sci.*, 312, 278, 1978.
93. **Jilek, F. and Hörmann, H.**, Fibronectin (cold-insoluble globulin), VI. Influence of heparin and hyaluronic acid on the binding of native collagen, *Hoppe-Seylers Z. Physiol. Chem.*, 360, 597, 1979.
94. **Yamada, K. M., Kennedy, D. W., Kimata, K., and Pratt, R. M.**, Characterization of fibronectin interactions with glycosaminoglycans and identification of active proteolytic fragments, *J. Biol. Chem.*, 255, 6055, 1980.
95. **Johansson, S. and Höök, M.**, Heparin enhances the rate of binding of fibronectin to collagen, *Biochem. J.*, 187, 521, 1980.
96. **Dessau, W., Sasse, J., Timpl, R., Jilek, F., and von der Mark, K.**, Synthesis and extracellular deposition of fibronectin in chondrocyte cultures, *J. Cell Biol.*, 79, 342, 1978.
97. **Gold, L. I. and Pearlstein, E.**, Fibronectin-collagen binding and requirement during cellular adhesion, *Biochem. J.*, 186, 551, 1980.
98. **Ginsberg, M. H., Painter, R. G., Birdwell, C., and Plow, E. F.**, The detection, immunofluorescent localization, and thrombin-induced release of human platelet-associated fibronectin antigen, *J. Supramol. Struct.*, 11, 167, 1979.
99. **Hynes, R. O., Ali, I. V., Destree, A. T., Mautner, V., Perkins, M. E., Senger, D. R., Wagner, D. D., and Smith, K. K.**, A large glycoprotein lost from the surfaces of transformed cells, *Ann N.Y. Acad. Sci.*, 312, 317, 1978.
100. **Arneson, M. A., Hammerschmidt, D. E., Furcht, L. T., and King, R. A.**, Fibronectin-correctable platelet function defect in Ehlers-Danlos syndrome, *JAMA*, 244, 144, 1980.
101. **Rubin, K., Höök, M., Öbrink, B., and Timpl, R.**, Substrate adhesion of rat hepatocytes: mechanism of attachment to collagen substrates, *Cell*, 24, 463, 1981.
102. **Rubin, K., Oldberg, A., Höök, M., and Öbrink, B.**, Adhesion of rat hepatocytes to collagen, *Exp. Cell Res.*, 117, 165, 1978.
103. **Stanley, J. R., Foidart, J. M., Murray, J. C., Martin, G. R., and Katz, S. I.**, The epidermal cell which selectively adheres to collagen substrates in the basal cell, *J. Invest. Dermatol.*, 74, 54, 1980.
104. **Schiffmann, E. and Gallin, J. I.**, Biochemistry of phagocyte chemotaxis, *Curr. Top. Cell Regul.*, 15, 203, 1979.
105. **Postlethwaite, A. E., Synderman, R., and Kang, A. H.**, Chemotactic attraction of human fibroblasts to a lymphocyte-derived factor, *J. Exp. Med.*, 144, 1188, 1976.
106. **Ali, I. U. and Hynes, R. O.**, Effect of LETS glycoprotein on cell motility, *Cell*, 14, 439, 1978.
107. **Seppä, H., Grotendorst, G., Seppä, S., Schiffmann, E., and Martin, G. R.**, The platelet-derived growth factor is a chemoattractant for fibroblasts, *J Cell. Biol.*, 92, 1981.
108. **Hewitt, A. T. and Varner, H. H.**, unpublished observations.
109. **Terranova, V. P.**, Unpublished.

Chapter 7

# THE IMMUNOBIOLOGY AND IMMUNOGENETICS OF THE COLLAGENS*

**John D. Kemp and Joseph A. Madri**

## TABLE OF CONTENTS



* Supported in part by National Cancer Institute Contract 1-CB-84255-37.

# I. ABSTRACT

This chapter serves as an extension of previous reviews on the immunobiology of collagen. The focus is initially upon H-2 linked IR gene phenomena in the control of murine antibody response to heterologous collagens. The discussion is subsequently directed towards the differences in cognitive specificities between T cells and B cells. Insights concerning these questions gained from studies of the immune response against the lysozymes are applied to the problems encountered in the immune response to collagen. Finally, the relevance of the discussion to the problems of monoclonal antibody production and autoimmune disease is mentioned.

# II. IR GENES AND COLLAGEN

The original work indicating that the murine antibody response to immunization with heterologous collagens was controlled by H-2 linked immune response (IR) genes can be attributed to Timpl's laboratory.[1–4] In careful studies using purified native Type I, Type II, and Type I procollagen from calves, they demonstrated several findings that are considered classic features of IR gene control. Upon immunization, strains of mice bearing particular H-2 haplotypes behaved uniformly and scored as either high, intermediate, or low IgG antibody producers. Response status did not change upon hyperimmunization. F1 hybrids between high and low responders behaved as high responders, and when such F1 mice were used for a first-generation backcross breeding to the low responder parent, high responsiveness was detected in half of the progeny. Thus, it appeared that a single dominant autosomal gene controlled for high responsiveness. This gene could be mapped to the IA subregion of the H-2 complex, and the particular alleles associated with high responsiveness to Types I and II calf collagen and the procollagen peptide of Type I collagen are shown in Table 1. To establish that these findings were dependent upon intact T cell function, they attempted immunization of nude mice of a high responder haplotype and found no antibody production.

Point for point, these observations fell in line with a large literature detailing the importance of H-2 linked IR genes in the control of immune responses to foreign thymus-dependent protein antigens[5] and suggested that the basic structure of the immune response to collagen was similar in most major respects to that seen for other diverse families of homologous mammalian proteins such as the lysozymes, insulins, and myoglobulins. These studies thus laid the groundwork for subsequent investigations into the genetic control of the immune response to other members of the collagen family and encouraged some groups to look upon collagen as a useful antigenic probe in basic studies of the immune response.[6,7]

Recently, our laboratory has been engaged in the study of the murine immune response to native Type III and Type V ($AB_2$) human collagen.[8] Initial test panels of a diverse group of inbred mice indicated dramatic strain-associated variability in antibody titers against both types of collagen. In both cases, it was possible to choose groups of H-2 recombinant inbred strains which would directly address the question of IR gene effects.

Insofar as Type III human collagen is concerned, the H-2-linked immune response control shows a basic similarity to that seen for Types I and II calf collagen and Type I procollagen (see Table 1). It would appear that a gene (or genes) mapping to the K, IA, or IB subregions is the key determinant of responsiveness. In particular, when B10 recombinant inbred lines that display various portions of the "s" haplotype are immunized, the high responder B10.A (9R) is the most informative. This strain possesses the "s" allele at K and at the IA subregion, as do the B10.S and B10.HTT. However, the alleles in the remainder of the H-2 complex in the B10.A (9R), from IJ to D, are otherwise associated with low responsiveness. The allele for IB is indeterminate in the B10.A (9R).

The data obtained where inbred mice were immunized with Type V collagen is shown in Table 1. Once again H-2-linked effects are clearly seen when one compares the low responder B10 to the high responder B10.A. The complexity of the effects is apparent when the B10.A (3R), B10.A (4R), and B10.A (5R) are all noted to be high responders. While the presence of the "k" allele at IA in the B10.A and the B10.A (4R) seems sufficient to explain their responses, the results seen with the B10.A (3R) and B10.A (5R) would seem to require the mechanism of two-gene complementation. A more detailed treatment of this topic is given in our recent paper.[8] Recent work by Hedrick and Watson also indicates that two gene complementation effects can be shown to occur when studying the antibody response to partially purified Type I calf skin collagen.[9]

Thus far, the data available support the idea that the collagens generally behave like classical thymus dependent antigens under IR gene control. The data also suggest (at minimum) that each type of collagen will display a different and characteristic pattern of high and low responder H-2 alleles. Furthermore, although sheep and bovine Type I procollagen exhibit some similarity in haplotype response patterns,[10,11] the information currently available does not guarantee that the same type of collagen, taken from more distant species, will show the same responder allele patterns in mice. Should related collagens from different species eventually be shown to differ in such a manner, then the results might be attributable to very small differences in primary amino acid sequence. This type of finding has been made and studied in some detail in the immune responses against other families of homologous antigens such as the lysozymes.[12]

The preceding discussion is meant to convey the practical importance of H-2-linked IR gene control of the immune response to the collagens. The acceptance of the importance of the IR gene effect, however, leads onto other questions. One set of questions has to do with the cell surface molecules involved in IR gene function. This is touched upon very briefly to provide some basis for the consideration of the complex observations discussed above, particularly the problem of two gene complementation. Subsequently, experiments dealing with regulatory interactions at the cellular level are discussed.

### A. Cell Surface Molecules and IR Genes

The most important site of H-2-linked IR gene function appears to be a membrane interface between a subset of T cells and the antigen presenting cell (APC).[13] The APC is a bone marrow monocyte-derived cell that expresses certain surface glycoproteins whose structural genes are coded for in the I region of the H-2 complex. The I region controls many immune phenomena and is composed of five subregions: IA, IB, IJ, IE, and IC. In one terminology, the two possible pairs of relevant expressed glycoproteins are called $A_\alpha$, $A_\beta$ and $A_e$, $E_\alpha$. This nomenclature derives from the surprising discovery that three of the four chains are coded for in the IA subregion ($A_\alpha$, $A_\beta$, and $A_e$), while only one ($E_\alpha$) is coded for in the IE subregion.[14] In some strains, the $A_e$ chains pairs up with the $E_\alpha$ chain to produce a surface pair that is structurally homologous to the $A_\alpha$, $A_\beta$ pair. Oddly enough, but importantly, other strains fail to produce the $E_\alpha$ chain and do not express the cell surface pair, $A_eE_\alpha$ (they do express $A_\alpha$, $A_\beta$). Such strain variations in cell surface glycoprotein production may account for some of the apparent complexity of two-gene comple-

**Table 1**
**SUMMARY OF THE RESPONDER STATE OF INBRED MOUSE STRAINS TO VARIOUS CALF, SHEEP, AND HUMAN PROCOLLAGEN AND COLLAGENS**

| Antigen | Mouse stain | H-2 haplotype | H-2 Complex regions K | A | B | J | E | C | S | G | D | Responder state | Ref. |
|---|---|---|---|---|---|---|---|---|---|---|---|---|---|
| Procollagen peptide (Col 1) of Type I collagen, sheep and calf skin | B10 | b | b | b | b | b | b | b | b | b | b | High | 10, 11 |
| | B10.BR | k | k | k | k | k | k | k | k | k | k | High | |
| | B10.A | $a_1$ | k | k | k | k | k | d | d | d | d | High | |
| | AQR | y | q | k | k | k | k | d | d | d | d | High | |
| | B10.HTT | $t_3$ | s | s | s | s | k | k | k | k | d | Intermediate | |
| | B10.D2 | d | d | d | d | d | d | d | d | d | d | Low | |
| | DBA/1 | q | q | q | q | q | q | q | q | q | q | Low | |
| | DBA/2 | d | d | d | d | d | d | d | d | d | d | Low | |
| | A.SW | s | s | s | s | s | s | s | s | s | s | Low | |
| Type I procollagen, sheep and calf skin | C57BL/10Sn | b | b | b | b | b | b | b | b | b | b | High | 1, 10, 11 |
| | B10.BR/SgSn | k | k | k | k | k | k | k | k | k | k | High | |
| | B10.D2/nSn | d | d | d | d | d | d | d | d | d | d | High | |
| | DBA/2 | d | d | d | d | d | d | d | d | d | d | High | |
| Acid-soluble Type I collagen, calf skin | C57B1/10 | b | b | b | b | b | b | b | b | b | b | High | 2 |
| | B10 | b | b | b | b | b | b | b | b | b | b | High | |
| | C3H.SW | b | b | b | b | b | b | b | b | b | b | High | |
| | ABY | b | b | b | b | b | b | b | b | b | b | High | |
| | LP | b | b | b | b | b | b | b | b | b | b | High | |
| | D1.LP | b | b | b | b | b | b | b | b | b | b | High | |
| | B10.M/Sn | f | f | f | f | f | f | f | f | f | f | High | |
| | A.CA | f | f | f | f | f | f | f | f | f | f | High | |
| | ASW | s | s | s | s | s | s | s | s | s | s | High | |
| | SJL | s | s | s | s | s | s | s | s | s | s | High | |
| | B10.BR/SgSn | k | k | k | k | k | k | k | k | k | k | Low | |
| | C3H | k | k | k | k | k | k | k | k | k | k | Low | |
| | AKR | k | k | k | k | k | k | k | k | k | k | Low | |

| | | | | | | | | | | | | | |
|---|---|---|---|---|---|---|---|---|---|---|---|---|---|
| | B10.D2/nSn | d | d | d | d | d | d | d | d | d | d | Low | |
| | DBA/2 | d | d | d | d | d | d | d | d | d | d | Low | |
| | NZB | d | d | d | d | d | d | d | d | d | d | Low | |
| | B10.A | a | a | a | a | a | a | a | a | a | a | Low | |
| | A | a | a | a | a | a | a | a | a | a | a | Low | |
| | B10.RIII (7INS)/sn | r | r | r | r | r | r | r | r | r | r | Low | |
| | LP.R III | r | r | r | r | r | r | r | r | r | r | Low | |
| | DBA/1 | q | q | q | q | q | q | q | q | q | q | Low | |
| | SWR | q | q | q | q | q | q | q | q | q | q | Low | |
| | B10.AKM/Sn | m | m | m | m | m | m | m | m | m | m | Low | |
| | AKR.M | m | m | m | m | m | m | m | m | m | m | Low | |
| | C3H.NB | p | p | p | p | p | p | p | p | p | p | Low | |
| | C3H.JK | j | j | j | j | j | j | j | j | j | j | Low | |
| | B10.PL | u | u | u | u | u | u | u | u | u | u | Low | |
| | B10.A 2R | hz | k | k | k | k | k | d | d | ? | b | Low | |
| | B10.A 4R | $h_4$ | k | k | k | b | b | b | b | b | b | Low | |
| | B10.A 5R | b | b | b | b | k | d | d | d | d | d | Low | |
| | A.TL | s | k | k | k | k | k | k | k | k | d | Low | |
| Pepsin-soluble Type II collagen, calf nasal septal | SJL/J | s | s | s | s | s | s | s | s | s | s | High | 4 |
| | DBA/1J | q | q | q | q | q | q | q | q | q | q | High | |
| | C57.BLc/10 | b | b | b | b | b | b | b | b | b | b | Low | |
| | B10.DR | k | k | k | k | k | k | k | k | k | k | Low | |
| | B10.D2 | d | d | d | d | d | d | d | d | d | d | Low | |
| Pepsin-soluble Type III collagen, human placental membranes | A/J | a | k | k | k | k | k | d | d | d | d | Low | 8 |
| | B10.A | a | k | k | k | k | k | d | d | d | d | Low | |
| | B10 | a | b | b | b | b | b | b | b | b | b | Low | |
| | 3R | $i_3$ | b | b | b | b | k | d | d | d | d | Low | |
| | 5R | $i_5$ | b | b | b | k | k | d | d | d | d | Low | |
| | Balb/c | d | d | d | d | d | d | d | d | d | d | Low | |
| | ABY | b | b | b | b | b | b | b | b | b | b | Low | |
| | 8R | $as_1$ | k | k | ? | ? | s | s | s | s | s | Intermediate | |
| | C57 | b | b | b | b | b | b | b | b | b | b | Intermediate | |
| | 4R | $h_4$ | k | k | b | b | b | b | b | b | b | Intermediate | |
| | DBA 1J | q | q | q | q | q | q | q | q | q | q | Intermediate | |
| | ASW | s | s | s | s | s | s | s | s | s | s | High | |
| | B10S | s | s | s | s | s | s | s | s | s | s | High | |
| | B10.HTT | $t_3$ | s | s | s | s | k | k | k | k | d | High | |
| | 9R | $t_4$ | s | s | ? | k | k | d | d | d | d | High | |

**Table 1 (Continued)**
**SUMMARY OF THE RESPONDER STATE OF INBRED MOUSE STRAINS TO VARIOUS CALF, SHEEP, AND HUMAN PROCOLLAGEN AND COLLAGENS**

| | | | H-2 Complex regions | | | | | | | | | | |
|---|---|---|---|---|---|---|---|---|---|---|---|---|---|
| **Antigen** | **Mouse stain** | **H-2 haplotype** | **K** | **A** | **B** | **J** | **E** | **C** | **S** | **G** | **D** | **Responder state** | **Ref.** |
| Pepsin-soluble Type V ($AB_2$) collagen, human placental membranes | B10.D2 | d | d | d | d | d | d | d | d | d | d | Low | 8 |
| | B10 | b | b | b | b | b | b | b | b | b | b | Low | |
| | C57B6 | b | b | b | b | b | b | b | b | b | b | Low | |
| | DBA.1J | q | q | q | q | q | q | q | q | q | q | Low | |
| | ASW | s | s | s | s | s | s | s | s | s | s | Intermediate | |
| | ABY | b | b | b | b | b | b | b | b | b | b | High | |
| | Balb/c | d | d | d | d | d | d | d | d | d | d | High | |
| | A/J | a | k | k | k | k | k | d | d | d | d | High | |
| | B10A | a | k | k | k | k | k | d | d | d | d | High | |
| | 4R | $h_4$ | k | k | d | d | d | d | d | d | d | High | |
| | 5R | $i_5$ | b | b | b | k | k | d | d | d | d | High | |
| | 3R | $i_3$ | b | b | b | b | k | d | d | d | d | High | |

mentation effects in general, and in particular they may account for the results seen in the response to Type V collagen (see Reference 8 and this paper).

It has become increasingly clear that one or more groups of T cells recognize and react vigorously to antigens displayed in close conjunction with the APC cell surface glycoproteins coded for in the I region of H-2, and that the highly polymorphic amino acid sequences of these glycoproteins are intimately associated with the IR gene phenomenon. In fact, it has been possible to selectively interfere with T cell-dependent antibody responses and antigen induced T cell proliferation assays with antibodies that recognize the I region glycoproteins on the APC.[15] A recent report has indicated that the study of the T cell response to collagen and collagen-like moieties as measured by the antigen dependent in vitro proliferation assay may be successfully performed.[7]

Given tools such as the in vitro T cell proliferation assay, it should be possible to map with great precision those portions or fragments of any type of collagen molecule that elicit certain T cell responses during immunization. Of course it goes almost without saying that the utmost care must always be taken in antigen purification and characterization prior to immunization in order to accurately interpret such experiments. Such data may then be compared and contrasted with the extensive literature available concerning the moieties that react with elicited anticollagen antibodies.[16] Inevitably, questions that have previously arisen in studies of other antigens will be critical in the comprehension of the data derived from studies of collagen. For example, how do the specificities of antibodies and T cells compare? A series of very impressive experiments has approached several of these questions in regard to the immunoregulation of the immune response to the lysozymes.

**B. The Lysozyme Model: Clues for Collagen?**

A recent set of studies concerning the regulation of the murine immune response to heterologous lysozymes has begun to shed light on the complexity of antigen perception by subsets of cells, particularly T cells, and will be of great value in the ultimate comprehension of the immunobiology of collagen. The experiments, performed in Sercarz' laboratory[17,18] have shown that the lysozyme molecule can be divided into two rough "immunologic" zones (see Figure 1). The zones are called "N-C" and "LII" and correspond to CNBr peptide fragments. One fragment contains the amino terminus and the carboxyl terminus of the molecule which are linked by a single disulfide bond. This is the "N-C" zone. The other fragment, "LII", contains the remainder of the molecule, within which there are three disulfide bonds. Their discoveries include the following:

1. Animals that are genetic responders to the lysozyme under study produce antibodies almost exclusively directed against the "NC" fragment
2. Furthermore, the responders have two requisite groups of T helper cells; one group, the "carrier-helpers", see determinants in the "LII" region, and another group, the "idiotypic helpers", help the B cells that see "N-C"
3. Some animals that are nonresponders to the lysozyme under study develop suppressor T cells that are directed, like B cells, against "N-C"

These animals apparently do not have the idiotypic helpers, but oddly enough are still capable of generating "carrier helpers" against "LII". These findings and others have begun to illuminate some of the significant differences in the cognitive realms of T cells and B cells where exposed to an antigen under IR gene control.

The relevance of this line of thought to collagen is much more clearly seen when

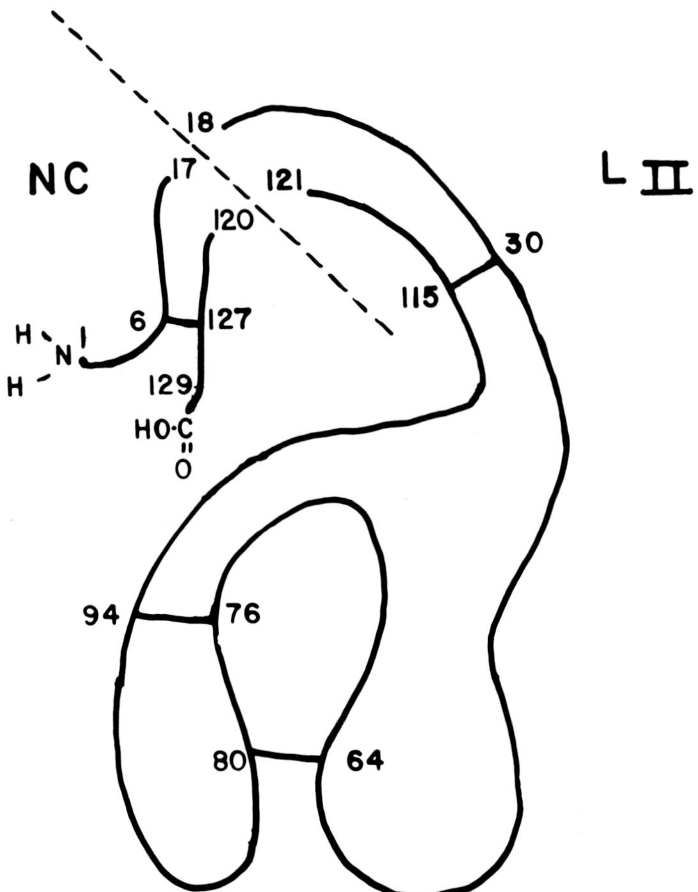


FIGURE 1. Structure of hen egg lysozyme molecule — amino acid residues 1 to 17 and 120 to 129 containing a Cys-Cys disulfide bond at residue numbers 6 and 127 represents the N-C peptide seen in the box labeled "NC", while amino acid residues 18 to 119 containing Cys-Cys disulfide bonds at residue numbers 30 and 115, 64 and 80, and 76 and 94 represent the LII peptide, labeled LII. (Redrawn from Adorini, L., Harvey, M., Miller, A., and Sercarz, E., *J. Exp. Med.*, 150, 293, 1979. With permission.)

the earlier paper by Rohde et al.[11] is recalled. During the analysis of the immunogenicity of sheep Type I procollagen, it became clear that "carrier-like" determinants were invested in a segment of the amino terminal CNBr peptide (CBO, 1) that consisted of a collagenous sequence (Col 3) and a short nonhelical region (Col 2). "Hapten-like" determinants, i.e., those determinants that reacted with elicited antibody mapped to a distinctly different part of the peptide (Col 1). A strong potential analogy exists therefore between the "carrier-like" region of CBO, 1 (Col 2 and 3) and the LII region of lysozyme (see Figure 2). Likewise the haptenic portion of

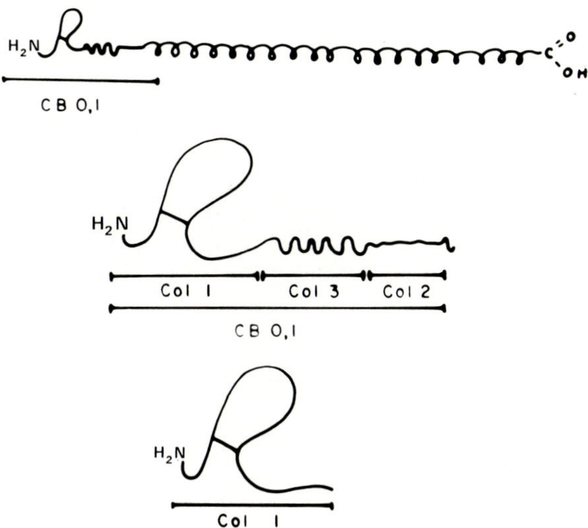


FIGURE 2. Structure of several procollagen antigens denotes the presence of five intrachain Cys-Cys disulfide bonds. Col 1 is individually demarcated as the Col 3, Col 2 fragments. (Redrawn from Rohde, H., Nowack, H., and Timpl, R., *Eur. J. Immunol.*, 8, 141, 1978. With permission.)

CBO, 1 (Col 1) may be analogous to the N-C region of lysozyme. The attendant implications for T cell reactivities and specificities follow directly.

A feature of the sheep procollagen experiments that is of special additional interest is the fact that the role of antigen conformation was examined. As one might expect, the antibody response was completely conformationally dependent. The carrier effect — most likely due to T helper cell activity—was, however, independent of conformation. Although detailed cellular analysis of these problems has not been performed in the lysozyme system, experiments in another antigen system have indicated that some T helpers are conformationally independent, while suppressor T cells, like antibodies, show conformational specificity.[19]

Apart from these questions, however, the role of suppression in immunoregulation in the lysozyme system has been intensively studied and, as noted earlier, some features of the specificities of suppressor T cells have been elucidated. Only one paper, that of Nowack et al.,[10] has even touched upon the possibilities of suppressor effects in the regulation of the immune response to bovine Type 1 collagen. In their study, it was observed that some mice were low responders to collagen, but high responders to procollagen. Furthermore, when such mice were preimmunized with collagen, a subsequent challenge with procollagen yielded very low antibody titers against procollagen. Similar effects have been seen with the synthetic terpolymer GAT[5] and have been shown to be attributable to the development of suppressor T cells. Although Nowack et al. did not make such a claim to account for their data, the current authors feel that such a possibility is strong indeed. It is even more intriguing to note that the work by Rohde et al., discussed earlier, appears to have arisen, in part, as a result of the questions raised by Nowack et al. The same strains of mice used in the bovine Type I procollagen experiments were evaluated for their responses to sheep Type I collagen, and similar response patterns were observed. If suppressor T cells do account for some of the data observed by Nowack et al., and if the data on B cell and helper T cell specificities obtained by Rohde et al. may be jointly considered, then overall models very similar to those generated for lysozyme immunoregulation may be applicable to collagen.

As a form of summary, it should be pointed out that so far, the collagens have given every indication that they represent an accessible and important group of molecules that may be used in basic immunologic experiments.[6,7] Unpublished recent experiments from our laboratory have shown that Type I collagen can be easily conjugated with simple haptens such as TNP and that when immunized with TNP-collagen conjugates, mice that are responders to collagen form rapid, high-titer anti-TNP antibody, while nonresponders to collagen fail to do so.

### C. Immediate Practical Implications for Collagen Studies

As mentioned in the introduction to this article, one of the pressing current needs is for the production of specific, well-characterized monoclonal antibodies against the collagens. Part of the value of discussion of IR genes is based partly upon the idea that one is more likely to achieve this goal by using animals that are genetically capable of producing high titered antibodies against the antigen of interest. It seems clear that individualized strategies will be necessities for each type of collagen, and indeed may be required for each component part of a collagen molecule. The interesting additional possibility that mice with widely divergent background genes (other than H-2) may respond to various groups of determinants on the same molecule should not be forgotten in designing protocols for monoclonal antibody production.

Finally, the current and future studies of IR gene phenomena and T and B cell cognitive specificities in relation to collagen will undoubtedly have an eventual impact on the persistent issue of the autoantigenic nature of the collagens. Collagen has been implicated in one way or another, in experimental models of arthritis[20] and pulmonary fibrosis.[21] It should be kept in mind that H-2-linked effects are now well documented in several experimental autoimmune syndromes such as myasthenia gravis[22] as well as in clinical situations such as ankylosing spondylitis.[23] Just as a very small portion of the myelin basic protein can induce full-blown experimental autoallergic encephalomyelitis,[24] so it may be that tiny portions of a given type of collagen molecule may interact with IR gene products under certain circumstances so as to stimulate an autoimmune disease. Clearly, much more careful work needs to be done before any definite statements can be made.

## REFERENCES

1. **Hahn, E., Nowack, H., Goetze, D., and Timpl, R.,** H-2 linked genetic control of antibody responses to soluble calf skin collagen in mice, *Eur. J. Immunol.,* 5, 288, 1975.
2. **Nowack, H., Hahn, E., David, C., Timpl, R., and Gotze, D.,** Immune response to calf collagen type 1 in mice: a combined control of IR-1A and non H-2 linked genes, *Immunogenetics,* 2, 331, 1975.
3. **Nowack, H., Hahn, E., and Timpl, R.,** Requirement for T cells in the antibody response of mice to calf skin collagen, *Immunology,* 30, 29, 1976.
4. **Nowack, H., Hahn, E., and Timpl, R.,** Specificity of the antibody response in Inbred mice to Bovine Type I and Type II collagen, *Immunology,* 29, 621, 1975.

5. **Benacerraf, B. and Germain, R.,** The immune response genes of the major histocompatibility complex, *Immunol. Rev.,* 38, 70, 1978.
6. **Hedrick, S. and Watson, J.,** Genetic control of the immune response to collagen II. Antibody responses produced in fetal liver restored radiation chimeras and thymus reconstituted $F_1$ hybrid nude mice, *J. Exp. Med.,* 150, 646, 1979.
7. **Rosenwasser, L., Bhatnagar, R., and Stobo, J.,** Genetic control of the murine T lymphocyte proliferative response to collagen: analysis of the molecular and cellular contributions to immunogenicity, *J. Immunol.,* 124, 2854, 1980.
8. **Kemp, J. and Madri, J.,** The immune response to human type III and type V ($AB_2$) collagen: antigenic determinants and genetic control in mice, *Eur. J. Immunol.,* 11, 90, 1981.
9. **Hedrick, S. and Watson, J.,** Genetic control of the immune response to collagen I. Quantitative determinations of response levels by multiple I-region genes, *J. Immunogenet.,* 7, 271, 1980.
10. **Nowack, H., Rohde, H., Gotze, R., and Timpl, R.,** Genetic control and carrier and suppressor effects in the antibody response of mice to procollagen, *Immunogenetics,* 4, 117, 1977.
11. **Rohde, H., Nowack, H., and Timpl, R.,** Localization of antigenic activity and immunogenic capacity in different conformational domains of procollagen peptide, *Eur. J. Immunol.,* 8, 141, 1978.
12. **Secarz, E., Yowell, R., Turkin, D., Miller, A., Arraneo, B., and Adorini, L.,** Different functional specificity repertoires for suppressor and helper T cells, *Immunol. Rev.,* 39, 108, 1978.
13. **Rosenthal, A. S.,** Determinant selection and macrophage function in genetic control of the immune response, *Immunol. Rev.,* 40, 136, 1978.
14. **Jones, P., Murphy, D., and McDevitt, H.,** Two gene control of the expression of a murine IA antigen, *J. Exp. Med.,* 148, 925, 1978.
15. **Lerner, E., Matis, L., Janeway, C., Jones, P., Schwartz, R., and Murphy, D.,** Monoclonal antibody against an IR gene product?, *J. Exp. Med.,* 152, 1085, 1980.
16. **Timpl, R.,** *Immunological Studies on Collagen in Biochemistry of Collagen,* Ramachandran, G. N. and Reddi, A. H., Eds., Plenum Press, New York, 1976, 319.
17. **Adorini, L., Harvey, M., Miller, A., and Sercarz, E.,** Fine specificity of regulatory T cells. II. Suppressor and helper T cells are induced by different regions of hen egg-white lysozyme in a genetically non-responder mouse strain, *J. Exp. Med.,* 150, 293, 1979.
18. **Adorini, L., Harvey, M., and Sercarz, E.,** The fine specificity of regulatory T cells IV. Idiotypic complementarity and antigen-bridging interactions in the antilysozyme response, *Eur. J. Immunol.,* 9, 906, 1979.
19. **Endres, R. and Grey, H.,** Antigen recognition by T cells. I. Suppressor T cells fail to recognize cross-reactivity between native and denatured ovalbumin, *J. Immunol.,* 125, 1515, 1980.
20. **Trentham, D. E., Townes, A. S., and Kang, A. H.,** Autoimmunity to type II collagen: an experimental model of arthritis, *J. Exp. Med.,* 146, 857, 1977.
21. **Kravis, T. C., Ahmed, A., Brown, T. E., Fulmer, J. D., and Crystal, R. G.,** Pathogenic mechanisms in pulmonary fibrosis: collagen-induced migration inhibition factor production and cytotoxicity mediated by lymphocytes, *J. Clin. Invest.,* 58, 1223, 1976.
22. **Christadoss, P., Lennon, V., and David, C.,** Genetic control of experimental autoimmune myasthenia gravis in mice, *J. Immunol.,* 123, 2540, 1979.
23. **Brewerton, D., Caffrey, M., Hart, F., James, D., Nicholls, A., and Sturrock, R.,** Ankylosing spondylitis and HL-A27, *Lancet,* I, 904, 1973.
24. **Hashim, G.,** Myelin basic protein: structure, function, and antigenic determinants, *Immunol. Rev.,* 39, 60, 1978.

# INDEX

## A



## B

## D



## E



## F

## G



## H

# I

## K



## L



## M



## N

## O



## P

## R



## S

## T



## U



## V



## W



## X